INTEGRATING TECHNOLOGY INTO THE SCIENCE CURRICULUM

PRIMARY

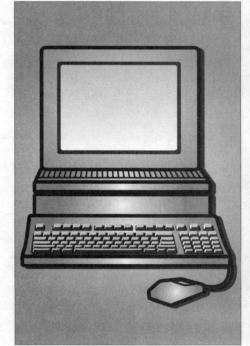

Illustrator:
Agi Palinay

Editor:
Evan D. Forbes, M.S. Ed.

Managing Editor:
Charles Payne, M.F.A., M.A.

Editor in Chief:
Sharon Coan, M.S. Ed.

Art Director:
Elayne Roberts

Art Coordination Assistant:
Cheri Macoubrie Wilson

Cover Artist:
Tina DeLeon

Product Manager:
Phil Garcia

Imaging:
Evan D. Forbes, M.S. Ed.

Acknowledgements:
Kid Pix Studio®, Copyright Brøderbund Software, Inc., 1996. All Rights Reserved.

Publishers:
Rachelle Cracchiolo, M.S. Ed.
Mary Dupuy Smith, M.S. Ed.

Author:
Diane Donato

Teacher Created Materials, Inc.
6421 Industry Way
Westminster, CA   92683
ISBN-1-57690-427-X

©1998 Teacher Created Materials, Inc.   Made in U.S.A.

The classroom teacher may reproduce copies of materials in this book for classroom use only. The reproduction of any part for an entire school or school system is strictly prohibited. No part of this publication may be transmitted, stored, or recorded in any form without written permission from the publisher.

TABLE OF CONTENTS

Introduction 4

Computer Access
- The One-Computer Classroom 5
- The Multi-Computer Classroom 8
- The Computer Lab 11

Assessment 14

Internet 15

Grant Writing 17

Integrated Lessons Plans

Health and Safety
- Self-Awareness 20
- Senses 22
- Safety 26
- Nutrition 29
- Living Things 33

Plants
- Plants Are Living Things 37
- The Parts of a Plant 40
- What Do Plants Need to Grow? 44

Animals
- Animals Are Living Things 47
- Ordering Animals 50
- Animal Groups 53
- The Parts of an Insect 59
- Animal Habitats 62

Interdependence
- Food Webs 66
- Predator/Prey Fantasy 69

TABLE OF CONTENTS (cont.)

Cycles
- The Plant Cycle 73
- The Life Cycle of a Mealworm Beetle 78
- Day and Night Sky 80
- Shadows 82
- Constellations and Myths 85
- The Moon's Cycle 87
- The Cycle of the Seasons 91
- The Water Cycle 96
- What's the Weather? 100
- Air Temperature 103

Changes
- Inside Earth 108
- Past, Present, Future 110
- Rocks Word Search 115
- Dinosaur Database 119
- Matching Fossils 124
- Swirling Solutions 126

Energy
- Heat, Light, and Sound 129
- Machines Help 135

Educational Software Distributors 137

Productivity Software 138

Reference Software 139

Storyboards 142

Bibliography 144

INTRODUCTION

When I was first introduced to computers in 1980, I was taught basic programming. We were shown how to make a loop. We had to develop an entire page of directions in order to have the computer create one screen that might say, "The sum of 4 plus 2 equals 6." I developed a computer camp for children to learn this type of programming.

Wow, have we come a long way! After several years I stepped out of teaching, which gave me insight into what the business world really needs in regards to technology. It appears that the old "necessity is the mother of invention" still holds true. The technology industry creates what businesses, artists, engineers, etc., want and need.

Obviously, education must and does keep up. The wonderful thing about our students is that they have no fear. Whatever they are asked to try on a computer becomes an exciting challenge to them, and they end up knowing more about the program than we, their teachers, do. So, we simply get them started and off they go. Usually, my students end up developing projects that are more creative than any I could ever imagine.

When integrating technology into the curriculum, remember how you use the tape recorder, camcorder, and other mechanical devices. The computer becomes a tool like the others. I recall when using the typewriter how we dreaded retyping the document because errors were made. Now children can get on the word processor and actually enjoy editing. With multimedia programs, all subjects easily become integrated with technology.

In *Integrating Technology into the Science Curriculum*, you will find lesson plans that invite you to "jump right in." Some plans may need adjusting to accommodate student ability and computer knowledge. These lesson plans make up most of this manual; however, there are also tips on management—various computer setups, assessment, surfing the Net, and grant writing. In the grant writing segment, you will learn how to find the money to pay for all of this new technology. Each lesson lists suggestions for two types of software. But do not let these suggestions limit you. Call the software companies' 800 numbers (see pages 137–141) to receive catalogs. There are new software products developed every day.

- **Reference** software is specific to the topic. This is software that teaches a lesson about your topic or simply gives desired information. Most software companies will allow you to preview their software before purchasing, so take advantage of this. You cannot always be sure that the software is just what you want for your program until you actually try using it.

- **Productivity** software is actually the tool used to create your finished products. For creating multimedia projects, you will want word processing, spreadsheets, database, draw/paint, and slide show capability. Find the programs that work best for you. I will be giving several different suggestions with each lesson plan, but if you haven't already, you will soon decide which productivity software you like best.

THE ONE-COMPUTER CLASSROOM

The one-computer classroom is a challenge for the teacher, but with some planning and thought it is manageable. While lessons will take longer to complete, sometimes days longer than with a lab, most lessons can be taught in the one-computer classroom.

Integration lessons, such as the ones found in this book, allow teachers to fit computer activities into lessons in other subject areas. This integration approach will help you find related tasks and activities for students who are not utilizing the computer. Computer work must fit with other tasks, and you must think about what other students will be doing while a few are at the computer or while you are at the computer with them. While this is not anything new to primary grade teachers, the computer does present some unique situations.

Students will have to be shown how to do certain tasks. These are directions of a physical nature, and at times they need to be step by step. This means you may have to spend some individual time with each student until each learns how to click the mouse, how to click on buttons and how to drag and drop objects on the screen. Some planning will be required to assure that you can work with one or two students while the rest of the class is busy at some other task.

A system for students to have time to work on assignments must be planned as well. One of the best methods is to rotate students on a daily basis to do assigned tasks, but if a projection device is available, the whole class can work together on some things.

Projection devices can be expensive but are worth the investment. Since a projection device is not something that you would use every day, it might make buying one more feasible if several teachers in your school requested one that could be shared. A projection device will allow you to show your whole class the computer screen at one time and allow you to introduce new material on the computer efficiently with one computer. Students will still need some one-on-one instruction with new skills, but their learning curve will be accelerated. Also, many of the lessons in this book can become whole class activities with a projection device for your computer.

There are several types of projection devices, and they come in a wide range of prices. Liquid-crystal display (LCD) panels are among the most affordable. They generally start around $1,500 for one with good resolution. They can cost up to more than $5,000 for deluxe models. LCD panels fit on top of an overhead projector and connect to your computer. The overhead projects the computer screen that is on the LCD panel. It is important to have an overhead that lights up from the base so that its light travels through the LCD panel. Most overheads sold today are made to do this. Older overhead projectors may have their light sources in the head that is above the base. This type of overhead will not work with an LCD panel.

Another popular type of projection device is a TV that is modified with a translator card and a special cord so that it can translate the computer's digital signal into an analogue signal that can be shown by the TV. This type of projection device usually costs less than an LCD panel but does not offer the clarity of picture or resolution found with LCD panels.

Projectors offer the best looking picture around, but they also are the most costly. Projectors usually start around $5,000. Projectors are one-piece devices that connect to the computer and then directly project onto the screen.

Computer Access

THE ONE-COMPUTER CLASSROOM (cont.)

For all prices, it is best to do some shopping in some catalogs. Check with area computer stores and the school librarian for catalogs that feature projection devices. These devices are becoming more affordable as they become more available, but they are not in everyone's budget.

If you have not bought your computer yet but know you are going to get only one, try to get an oversized monitor and arrange the system so that the monitor is slightly raised. This will allow a group of students to view the monitor as they sit around the computer table. While this solution is not a projection system, it will help make things a little easier when you are working with a small group around the computer.

You will spend some individual time with students in the beginning, but as students begin to acquire the basic skills needed to do lessons, they will become more independent on the computer, allowing you to work in other areas of the classroom. Students who use the computer regularly and see teachers using it regularly will develop a higher level of computer literacy more quickly than students who rarely use one or see it used. It is important that the computer be used for classroom teaching and as a tool for student use.

There are many related activities that begin with or follow up a lesson on the computer. Much of the curriculum for computers at the primary level involves teaching students the vocabulary of computers. This means giving them activities that promote intuitive understanding of what the terminology means. Students need to know what is actually happening when you direct a database software to sort or order records. With intuitive understanding of the functions, they will be able to interpret and use the results of the computer's response.

You can use this need to understand the meaning of computer vocabulary to keep students busy with physical manipulative activities that promote this understanding. This will mean that they are involved in related meaningful activity while you are working with a few students on basic computer skills.

Using the computer, talking about the computer, and demonstrating through experience what can be done with the computer will go a long way toward satisfying nearly any primary technology curriculum. It will also provide a sound understanding of the basics of using the computer to do everyday work that will help students to advance in both technical skills and understanding of concepts as they proceed through the grades with computers.

You do not have to have a lot of peripherals or a huge computer lab to teach computers. One computer with a paint program, multimedia software, and an integrated office package that provides database, spreadsheet, and word processing can be the basis for both satisfying your curriculum's objectives and your students' needs for the future when it comes to technology education.

The following is a suggested schedule for the one-computer classroom of approximately 25 students. Using the projected screen, you can demonstrate one or two activities to the students each week and then allow each student twenty minutes at the computer with a "buddy." When the "buddy" gets a turn, the other student then becomes the helper. In this way, each student is actually receiving forty minutes of computer time per week.

Computer Access

THE ONE-COMPUTER CLASSROOM (cont.)

By integrating computer time into the language arts block, the students waiting for a turn will be writing, reading, practicing cursive, doing science/social studies research, etc.

	Monday	Tuesday	Wednesday	Thursday	Friday
10:00	Red A	Blue A	Green A	Yellow A	Purple A
10:20	Blue B	Green B	Yellow B	Purple B	Red B
10:40	Green C	Yellow C	Purple C	Red C	Blue C
11:00	Yellow D	Purple D	Red D	Blue D	Green D
11:20	Purple E	Red E	Blue E	Green E	Yellow E

Possible Buddies: Red A/Purple C, Blue A/Red C, Green A/Blue C, Yellow A/Yellow D Purple A/Purple D, Blue B/Red D, Green B/Blue D, Yellow B/Green D, Purple B/Purple E, Red B/Red E, Green C/Blue E, Yellow C/Green E, and Yellow E alone or with another duo. Below are red and purple cards laminated together with letters to designate buddies. These same cards may be used all year, adding and deleting when new students arrive.

Sample of Buddy Cards

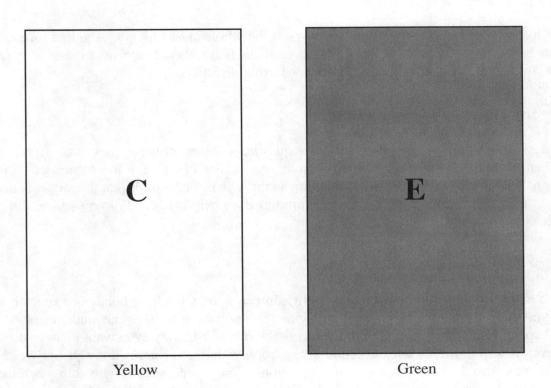

© Teacher Created Materials, Inc. 7 #2427 Integrating Technology into the Science Curriculum

Computer Access

THE MULTI-COMPUTER CLASSROOM

A multi-computer classroom has three or more computers. The multi-computer classroom allows the teacher to work with small groups or the entire class on technology-based activities or to use the computers as learning or research centers.

Noise Level Control

Many of the activities in a multi-computer classroom are cooperative learning based. In this environment, the students are encouraged to communicate with each other. A classroom of 25 students can create a lot of noise when they are working on a project! Therefore, it is necessary to have a system of noise level control. A numerical system works well for this. A 0 signifies silence; a 1, quiet; a 2, conversation; and a 3, presentation.

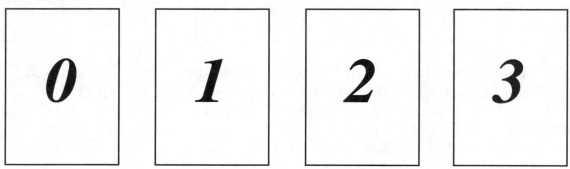

Laminated cards can be displayed at the computer or in the front of the classroom to indicate the acceptable noise level for an activity.

At times, it may be necessary to get the attention of the whole class quickly. A signal for complete silence could simply be to turn off the classroom lights. This tells the students to stop whatever they are doing, to look at the teacher, and to listen for further directions.

Asking for Help

As students work in their groups, they will have questions. Many of these questions can be answered by the other group members, but some will require the teacher's help. For these questions, the students need a signal. Use a brightly colored plastic cup as the signal for help. When the students need help, they place their cups on top of the computer monitor. If no help is needed, it remains on the computer table next to the monitor.

Behavior Management

Due to the nature of the cooperative learning environment, you must have behavior expectations and specific guidelines for acceptable student behavior. You will need to show the students exactly what it looks and sounds like when students are following the rules. You may even want them to do some role-playing exercises to reinforce what acceptable classroom behavior is and is not. I usually establish classroom rules with the students' input of suggestions. The students take ownership when they have helped established the rules, so I do this the first few days of the school year. A periodic review of "their" rules helps with behavior management.

THE MULTI-COMPUTER CLASSROOM (cont.)

What if a student chooses not to follow a rule? Then, you must have some very clear consequences for breaking the rule. It is important that you are consistent. Some sample consequences:

First Offense: warning

Second Offense: spend five minutes in time-out area

Third Offense: spend remaining time for activity in time-out area

Grouping Your Students

In a multi-computer classroom, it is necessary to organize the students into cooperative groups so that they can get the most out of computer time. A group of students is assigned to each computer. These students do research, create projects, publish writing, and engage in instructional activities on this computer. Make sure that you combine students with different abilities, genders, and technology competencies. Switch groups every six weeks to give your students an opportunity to work with others in the classroom. Your groups will be able to learn from each other and use their individual strengths and talents within the group. Many of the lessons in this book work well in the multi-computer classroom. Each member of the group needs to be assigned a specific task for the duration of the activity. These tasks should be put on laminated cards, and students should choose a task card before an activity begins. There are many tasks that can be performed by the students. The choice of tasks is determined by the computer activity. Some tasks that are appropriate for a computer-based activity are the following:

Recorder—records group's data

Waste Management—controls recyclables and trash

Tracker—makes sure that group members stay on task

Supplier—gets and returns supplies for group

Presenter—presents group's findings to class

Text Writer—inputs text into the computer

Graphics Artist—inputs graphics items into the computer

Calculator—performs calculations on the computer

Graph Maker—makes the computer graphs

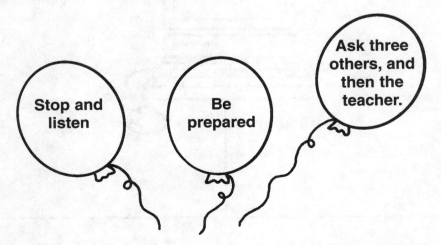

Computer Access

THE MULTI-COMPUTER CLASSROOM (cont.)

Computer Journals

It is helpful for each student to keep a computer journal. This journal can be written in at the end of each activity. The students can write about what they have learned about the subject, what they did on the computer, and whether they had any problems using the computer or working within the group. The journals allow the students to keep records of their progress and to express any problems that they may be experiencing. The journals can also be used by the teacher to track student progress, to resolve student problems, and to evaluate what the students have learned during an activity.

Equity

Making sure that all the students have an opportunity to work on the computer can be difficult. You can make this task easier if you use a visual display to indicate who has or has not been to the computer. Place a laminated square near each computer. Divide the square in half. The left side of the square is designated for those students who have not been to the computer, and the right side is for those students who have been. At the beginning of the week, place a clothespin for each group member on the "Not Been There Yet" side of the square. As the students participate in computer activities, they move their clothespins from the "Not Been There Yet" to the "Been There, Done That" side. When everyone in the group has had a turn, the clothespins are put on the "Not Been There Yet" side of the square, and the process is repeated.

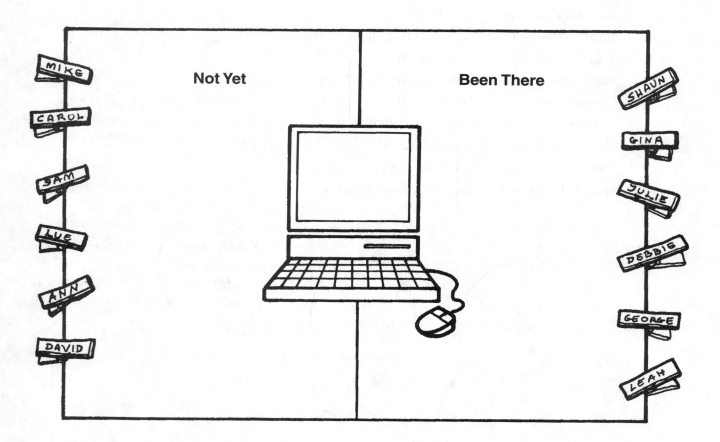

THE COMPUTER LAB

In one school, there is a computer lab with twenty-seven computers and over 700 students. Humanities and science teachers are paired with 50 students in heterogeneous groups. There are six math teachers, as well. There is a technology specialist who is in the lab as needed.

Each humanities and science class is assigned one class period per week to come to the computer lab. The humanities and science teachers may decide that for one week the humanities teacher will bring both classes to the computer lab and vice versa. The classroom teacher has the option of teaching the lesson, having the technology specialist teach the lesson, or having the technology specialist be in the lab for support. The classroom teacher must stay in the computer lab with the class. Activities during the time the class is in the computer lab must be curriculum related. This is not free time to play games or just draw pictures!

The lab schedule is set so that each class comes to the computer lab on Mondays, Tuesdays, Thursdays, or Fridays. Wednesdays are "open sign-up." That means that math teachers may decide to bring their classes to the lab on Wednesdays. Wednesdays also allow classroom teachers the flexibility to bring their classes to the lab for an additional class period during the week. This is particularly helpful when long-term projects are being developed.

The technology specialist is able to work with small groups of students in classroom mini-labs on Wednesdays. In addition to the computer lab computers, there are also 20 mobile computers (computers on carts) that are "stored" in classrooms. Stored computers are those that are placed in classrooms (approximately one for every pair of teachers). The stored computer is in the classroom to be used by the classroom teacher and the class unless they are needed by another teacher in order to set up a mini-lab. Any classroom teacher may request a mini-lab in his or her classroom for a period of time in order to work on technology projects in the classroom.

The technology specialist coordinates the "borrowing" of stored computers in order to create the mini-lab. The teacher requesting the mini-lab must pick up and return the stored computers. Setting up a database for mobile computers works well. The original location of the stored computers is listed, as well as the dates it has been borrowed. Going through the entire database once before borrowing a stored computer for the second time seems to be the fairest way to borrow the equipment.

The technology specialist develops the computer-lab schedule. The schedule is kept in the computer lab in a notebook. Classroom teachers must sign up on the schedule for the weekly activity and indicate how the technology specialist is needed for the weekly activity. This should be done three days in advance so that the technology specialist can make arrangements with other classroom teachers if he or she will be out of the lab during a particular class period.

Computer Access

THE COMPUTER LAB (cont.)

A seating chart is a must for the lab. Students should have permanently assigned seats. A boy-girl arrangement works very well, although there may be groans from the students the first day. Once it is explained to students that the goal is to work in the lab, not to socialize with friends, the seating arrangement is generally accepted.

Weekly Schedule

TIME	MON	TUES	WED	THUR	FRI
9:00 – 9:50	JOHNSON	BARKER	MORSE	SMART	COHEN
9:50 – 10:40	CARTER	SMART	HARRIS	MEDFORD	TERRY
10:40 – 11:30	MEDFORD	TERRY	CHANG	HARRIS	DOBER
12:00 – 12:50	DOBER	COHEN	JOHNSON	CARTER	GREEN
12:50 – 1:40	BARKER	MORSE	GREEN	CHANG	

Computer Access

THE COMPUTER LAB (cont.)

Seating Arrangement

Technology Specialist

Richard	Elena	Teacher	Jessica	Shawn
Molly	Michael	Sam	Alekhya	Jamie
Shane	Sarah	Taylor	Amanda	Matthew
Morgan	Cody	Kelly	Austin	Hannah
Travis	Ashley	Augusta	Alex	David

For free and independent use

© Teacher Created Materials, Inc. 13 #2427 Integrating Technology into the Science Curriculum

ASSESSMENT

Gone are the days of only paper and pencil multiple-choice tests used for assessment. With the understanding of various learning styles, teachers use a variety of tools to assess their students. By integrating technology into the primary curriculum, we are adding another tool by which to gauge our students' performances. Assessment of the students' work using computers is similar to assessment of any subject. Before beginning the assignment, the student must have a clear picture of what is expected. Consequently, part of creating our lesson plans is deciding how to measure the students' achievement. As the teacher explains the assignment, the student formulates a plan to achieve what is expected. Many suggestions for assessment are included with the integrated lesson plans in this manual.

Self-assessment helps the student to determine what went into the project. Often primary age children are very hard on themselves when doing a self-evaluation of the work completed, but it is important for the teacher to include the student's self-assessment as part of the overall grade.

Electronic portfolios, whereby the student creates slides demonstrating samples of his/her work using *Kid Pix*, *HyperStudio*, or other multimedia software, can begin in kindergarten and continue through twelfth grade. An annual electronic portfolio can illustrate work completed during one school year. I believe this is where we are going with technology in education. There are portfolio software programs, but you can begin now by using the above-mentioned multimedia software.

Teachers and students are held accountable for their achievements, and using the computer is an efficient, timesaving method for recordkeeping. Become familiar with a spreadsheet so that you can organize your students' grades and add comments when appropriate. Allow the students access to this ongoing assessment of their work.

INTERNET

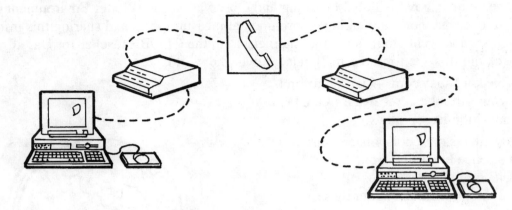

Where did all of this Internet communication begin? The military needed some way to know what was going on in all parts of the world. When our national defense is at stake, the sky's the limit, as you know.

Then college students started figuring out how to hook into this military network. Business and then private networks were not far behind. What a great way to reach countless numbers of potential customers. The military has since signed off on this public network and developed its own network. "What does this have to do with primary grade students?" you ask.

Not only will the children be able to surf the Net to gather information and enjoy learning, but there are numerous teacher sites as well. I will list several interesting URLs (Uniform Resource Locators or Internet addresses) for you to visit. But first, try doing your own search for information, using one of the search engines listed here:

    Alta Vista: http://www.altavista.digital.com
    Excite: http://www.excite.com
    InfoSeek: http://www2.Infoseek.com
    Lycos: http://www.lycos.com
    WebCrawler: http://webcrawler.com
    Yahoo!: http://www.yahoo.com

Tip: When searching more than one word, e.g., science+curriculum, connect your words with the plus sign. Otherwise, you will get everything found for science and everything found in curriculum. You need to be very exact when typing in a URL.

I will list only science URLs here, since we are discussing integrating technology into the science curriculum. However, there are places to visit for every topic imaginable.

- http://www.projectneat.org—National Education Advancement Team is an independent non-profit organization, with no political or religious affiliation, whose goal is to donate an Internet appliance to every K–12 school in the world. Sign on to see if any school in your district is registered. If not, you can receive the Internet appliance for your school. (Actually, this is not a science site, but hey, let's get the free hardware first!)

INTERNET (cont.)

- http://globe.fsl.noaa.gov/—Global Learning and Observation for a Better Environment (GLOBE) students all over the world are taking environmental measurements and sharing this information with scientists and technologists via the Internet. I am the GLOBE teacher for Dare County, N.C.; check out this site to find e-mail pals all over the world.
- http://www.dole5aday.com—Sign on to find out how your school can receive a free CD for your nutrition lesson plans.
- http://www.4Kids.org/coolspots/creature—This is a science hyperlink from www.4Kids.org which contains all subjects. Check out the information about these amazing animals:
- http://www.selu.com/~bio/PrimateGallery—Ape on the Web allows you to click on the animation of brachiating (the way some primates travel through the trees) or listen to the wild sounds of some primate vocalization.
- http://cvs.anu.edu.au/andy/beye/beye home.html—This site allows you to see the world through the eyes of a honeybee.
- http://www.bestfriends.com/main.htm—This site helps you and your pet live happily together.

Scavenger hunts are great opportunities to familiarize the students with the Internet. They can work with partners to locate information you request by using bookmarks. Once you find a site that you like, simply click on Bookmark and then Add bookmark to add it to a list of URLs. Ask questions that the children can answer by checking out the sites that you have already bookmarked. Now you are ready to explore the Internet. Have fun! But, be careful; the Internet is a great place to lose track of time.

Once you have surfed the Net to find more and more wonderful sites for both you and your students, you will want to share your ideas and do some networking of your own via e-mail. Check out the many services available. Free e-mail is available through http://www.juno.com. This does not give access to the Internet but is a free e-mail service. It might be all you want for now.

GRANT WRITING

It's the '90s and everyone uses technology, right? Maybe, but how much, how often, and why? The answer to these questions is determined by the big, bad budget! When meeting teachers from around the country, or even from your own state, I'm certain that you find the technology capabilities of their schools vary. Some schools are equipped with a full computer lab of 30 up-to-the-minute computers, many color printers, scanners, digital cameras, LCD panels for viewing—you name it, the works! These schools might also have several computer stations in each classroom, again fully equipped. On the other end of the spectrum are schools that might have a computer station available for two or three classrooms to share. Whatever the situation, we all want more! We have begged our PTAs, who continually do whatever they can for our children's needs; we have built business partnerships with many businesses in our community. It is time to turn to grant writing. Try to enlist the help of others on your staff, possibly creating a "grant team." It's always better to brainstorm with others on projects this important.

Who?

Begin with your school district. Many districts give small grants to teachers or teams of teachers for specific needs. They will offer stipends for projects developed that will enhance teaching and learning. Usually, these projects need to help more than one class and can be made available to other teachers. This means the project must be able to be replicated.

Next, seek out a community foundation which funds projects just in your county. Often RFPs (requests for proposals) are accepted quarterly, annually, etc. Many corporations are searching for places to put their money. Why? They not only are nice people—giving grants is a big tax write-off. If these corporations do not find ways to write off profits, Uncle Sam gets them. And speaking of Uncle Sam, the government also gives grants. There are both state and national grant opportunities.

Where?

"OK, I know who to ask, but where do I look to find these grant possibilities?" Actually, most libraries will have an index to help locate national and state grants. Corporations like *Kodak*® and *Coca-cola*® are also funders. The Internet is a wonderful source for addresses of grant makers. Of course, you might be seeking a grant because you need a computer with Internet access. Most libraries have Internet access. Here are a few tips when searching the Net: be specific and accurate; try grants + technology to connect those two words for the search engines.

GRANT WRITING (cont.)

What?

Now that you know who gives grants and where these persons are located, you will need tips on writing the grant.

- **State the Problem**

You cannot ask for money—it doesn't work that way. You must identify a need that can be fixed with technology. For example, "In order to increase third-grade reading scores by stimulating motivation to read, the children will create multimedia presentations after completing research on various third-grade science topics: earth/moon cycles, the food chain, animal habitats, etc."

- **Write the Plan**

Be very specific when telling exactly how you plan to improve reading scores with this strategy. "The student will choose an animal of choice, read many books about this animal, take notes while researching, and write paragraphs about the main topics. The student will create a multimedia slide show with the information garnered. The slide show will include clip art scanned from personal drawings of the animal and scrolling text boxes telling about the animal. The motivation generated by being able to use these hands-on materials will encourage the student to read, increase vocabulary, and demonstrate improved synthesizing, etc."

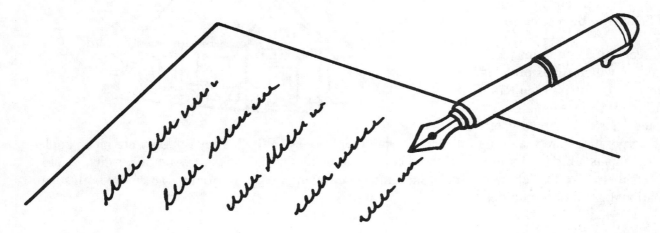

- **Prove It**

The folks who are reviewing the grant want hard evidence that you will attempt to measure the outcome of this plan. We must have evidence that technology can help solve the problem. "Seventy percent of the third-grade students will show at least a four-point improvement in the EOG reading scores. The entire faculty will receive training in the use of the hardware and software necessary to achieve these goals, etc."

GRANT WRITING (cont.)

- **Determine a Time Line**

Tell how long this project will take. "Staff development will begin September and be successfully completed in eight weeks; students will begin work at multimedia work stations by November; their slide shows will be completed by February. They will share their slide shows with parents at an open house and with other classrooms after the PTA Open House. These students will then become tutors for students the following year."

- **Research the Budget**

It is extremely important to itemize absolutely everything you need for this project, including personnel for training, stipends for staff development, and all hardware and software. This is where you demonstrate the research that goes into your grant. If continuing funds are necessary, you need to stipulate where you will get those funds.

How

This is not as difficult as mountain climbing, but just as the climber chooses his/her niches carefully, it takes some thought and practice to match the project with the funder. You will need to carefully read the RFP and follow all instructions to the letter.

Success!

It is very unusual for a first-time grant writer to achieve success. You need to remember that "No" really means, "Not at this time." If you are fortunate enough to meet success the first time, you will be energized to try for one even bigger. If you were told that your grant was a good one but not selected at this time, try again with a different funder. Look over the RFP carefully, and try again. Go back to your team for more brainstorming. Perhaps your budget needs revising; maybe something can be eliminated or postponed for later.

Good luck. There's lots of money out there. Go get it!

Integrated Lesson Plans

SELF-AWARENESS

Diversity in a child's world helps the child to understand others. Look for similarities and differences among classmates to promote self-awareness.

Materials:

Senses Reference Software
- *On the Playground*

Productivity Programs
- *Kid Pix* or *Kid Pix Studio*
- *The Amazing Writing Machine*

Procedure:

Into: Before the Computer

- Call children who have something in common to come to the front of the class: all who are wearing tennis shoes, long-sleeved shirts, have brown hair, etc.
- When at least one half of the children are up front, ask everyone to determine why they have been chosen. "What is the same about all of the children who are standing at the front of the class?" Depending on age, some prompting may be necessary.
- In order to include those seated, ask them if they can determine why these particular children are up in front. "Can anyone seated justify coming up and joining the others?"
- Then have the children standing divide themselves into two groups by something they have in common, and again allow the audience to determine why they are divided. Allow anyone seated to come to the front if he/she "matches" either group.
- The teacher asks one of the two groups to be seated.
- The remaining group then divides again, based on a different classification.
- This sorting continues until most children have had an opportunity to take part in the activity.
- Prior to beginning this lesson, the children have had practice identifying squares, circles, and triangles and the colors red, blue, green.
- They have also practiced drawing these figures with *Kid Pix Studio*.

*Kid Pix Studio is used for this activity.

SELF-AWARENESS (cont.)

Through: At the Computer

- Using *Kid Pix Studio* the children draw different-size green squares.
- They draw red squares.
- They draw blue squares.
- They draw green, blue, and red circles.
- They draw green, blue, and red triangles.
- They print their work.
- The teacher has put three huge overlapping outlines of circles on the chalkboard or bulletin board. (Venn diagram)
- The children cut out their shapes that they have printed.
- The teacher puts a few in the Venn diagram, asking why they are placed as they are.
- After putting a few more of the children's shapes in the Venn diagram, he/she invites the children to try placing theirs.

See below for an illustration of the Venn diagram.

Beyond: Extra Activities

- Using *Kid Pix Studio* children use stamps and work with a partner to group similar stamps.
- Using *HyperStudio* children work with a partner to create a three-card stack:
    - Title Card: Sorting/Classification by
    - Second Card: Similar Shapes
    - Third Card: Similar Colors
- *HyperStudio* media library, objects drawn on *Kid Pix* and imported, or clip art from other software (*The Amazing Writing Machine*) can be imported for this stack.

Venn Diagram Illustration

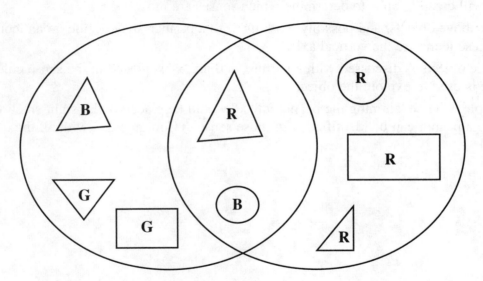

Integrated Lesson Plans

SENSES

Our senses help us to see, hear, smell, touch, and taste the world around us. We learn by using each or all of our senses. They keep us safe and allow us to observe everything as we go about our daily routines.

Materials:

The following are suggestions of items to arouse the senses; use as many of these as you like or others that you prefer: peanuts, coffee, sliced oranges, bananas, cologne, peanut butter, sliced lemons, sour/sweet pickles, sugar, salt, bells, alarm clocks, whistles, various cloths (including fake fur, sand, or sandpaper), glass/marble, cactus.

Productivity Programs
- *Kid Pix* or *Kid Pix Studio*
- *The Amazing Writing Machine*
- *The Graph Club*

Procedure:

Into: Before the Computer

- Utilize a variety of objects/strategies with the class to encourage the use of their senses.
- Point to many, many objects in the room for sight.
- Sniff peanuts, coffee rinds, sliced oranges, bananas, and cologne for smell.
- Sample peanut butter, sliced lemons, sour/sweet pickles, sugar, and salt for taste.
- Listen to bells, alarm clocks, whistles, clapping, stamping of feet, and giggles for hearing.
- Feel fake fur, sand, sandpaper, glass/marble, cactus, and various cloths for touch.

Through: On the Computer

- Students will create graphs to determine which senses are most used.
- Using the above objects, and possibly working with a partner, they decide on an icon for each sense. These icons are the vertical axis.
- Each of the objects is discussed with a partner, and an "x" is placed in the sense column when that sense is used to explore the object.
- For example, when examining the peanut an "x" would be placed under sight smell touch and taste since a peanut can be identified with those senses. Continue this with all the objects.

Integrated Lesson Plans

SENSES *(cont.)*

Through: On the Computer *(cont.)*

A Senses Graph

	Sight	Smell	Taste	Hearing	Touch
Peanuts	X	X	X	X	X
Fur	X				X
Alarm Clock	X			X	X

Beyond: Extra Activities

- Use the blackline assessment on the next two pages for students to write about their senses.
- Using *The Amazing Writing Machine*, the students can work in groups to write stories about their senses. They can add clip art to their stories.
- Students publish their work, and it can be put up on a "senses" bulletin board along with their graphs.

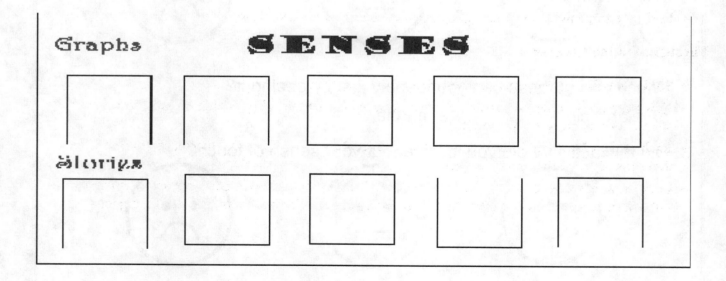

© Teacher Created Materials, Inc.

Integrated Lesson Plans

SENSES *(cont.)*
Assessment

Name _____

Directions: Tell if this activity worked well by circling the face that matches your feelings about what you did.

1. Did this activity help you learn about your sense of smell?

Which one of these can you identify just by smelling it?

a lemon　　　　　　　a marble　　　　　　　a jumping rope

2. Did this activity help you learn about your sense of sight?

Which one of these can you identify just by seeing it?

an alarm clock　　　　a marble　　　　　　　fur

3. Did this activity help you learn about your sense of touch?

Which one of these can you identify just by feeling it?

a lemon　　　　　　　glass　　　　　　　　fur

SENSES *(cont.)*

Assessment *(cont.)*

4. Did this activity help you learn about your sense of hearing?

   ☺ 😐 ☹

   Which one of these can you identify just by hearing it?

   a lemon   a marble   a bell

5. Did this activity help you learn about your sense of taste?

   ☺ 😐 ☹

   Which one of these can you identify just by tasting it?

   a lemon   an apple   a potato

   You might want to hold your nose while tasting an apple and a potato at home. I think you will be surprised.

   Write what you liked most about this activity.

   _____
   _____
   _____
   _____
   _____

Integrated Lesson Plans

SAFETY

Integrating technology, language arts, and science will be fun when the children develop a plan for one of the following safety practices and then use a draw program to create the three-dimensional artwork. This can be done independently or cooperatively as a jigsaw puzzle. When all of the safety practices are presented, the entire class will have enjoyed learning about a very important aspect of the science curriculum.

Materials:

- posters depicting safety practices in the home, in the classroom, on the playground, and on the school bus
- drawing paper
- crayons, markers, or paint
- stiff construction paper
- scissors
- glue, paste, or tape

Safety Reference Software

- *Firefighter*
- *Safety for Children: Playground Safety*

Productivity Programs

- *Microsoft Works*
- *Paint*
- *Kid Pix Studio*

Procedure:

Into: Before the Computer

- Use posters and pictures demonstrating safety practices at home and at school.
- If possible, allow time for the students to peruse any safety reference software available at the school.
- Discuss good safety practices.
- Either allow children to choose which safety practices to tell about, or put the children into safety groups to be sure all are involved.
- Allow time for the children to illustrate one of the safety practices discussed. Depending on age level, they can write a word, a sentence, or a paragraph about their illustration.

*For the example *Microsoft Works Word Processor* and *Paint* were used.

SAFETY *(cont.)*

Through: On the Computer

- Using their drawings as a guide, the children create their art with a paint program.
- They need to draw two or three different screens. For example, the first screen will be the background of trees, grass, and sky for the playground. If demonstrating safety at home, a background of a room with a window might work.
- The second screen will depict the people showing safety.
- A third screen with something in the foreground could be added, if desired.
- This computer art could show outlines only so that the children can color it after it is printed.
- After printing each screen, the children use their crayons or paint to add whatever they feel they need.
- Now they are ready to assemble their three-dimensional safety poster.
- Fold the stiff construction paper in half so that it can stand on its own.
- Cut out and paste on the background.
- Then cut out the main screen, fold the bottom under to help it stand up, and paste or tape it onto the other part of the construction paper.

Integrated Lesson Plans

SAFETY *(cont.)*

Through: On the Computer *(cont.)*

- If a third screen is used, color and paste it onto the construction paper a few inches in front of the main screen.
- Using a word processor, have the children each type a word, a sentence, or a paragraph about their safety project.
- Print it, cut it out to the desired size, and tape it onto the construction paper.

   Only two people at a time can safely play on the seesaw.

- Now the children are ready to set up their safety corner. After presenting their three-dimensional safety projects, the children will put them on a table in the classroom with a sign:

   We Know How to Be Safe

Beyond: Extra Activities

- Bring in community helpers to tell about safety; take digital snapshots of these helpers; print out the snapshots and paste them onto cardboard to add to the three-dimensional projects.
- Set up the three-dimensional safety projects outside the classroom and invite other classes to visit and learn about safety.
- Use this three-dimensional project strategy for other science activities: Care of Pets, Our Solar System, What's the Weather?, All About Me, Our Senses, etc.

Integrated Lesson Plans

NUTRITION

Children learn to like and dislike various foods at an early age. It is important for them to understand the value of eating a wide variety of foods from the food pyramid in order to have healthy bones, teeth, skin, muscles, eyes, etc. A motivational hook for this activity would be to promise a "nutritious feast" as a culmination of the understanding of the food pyramid. It might be a good idea to begin this lesson prior to the Thanksgiving or Christmas break.

Materials:

- food pyramid chart
- white paper plates
- crayons
- overhead projector and transparency (See page 32 for blackline transparency master.)

Nutrition Reference Software

- *5 A Day Adventures* (This CD is free of charge to teachers by writing to: *5 A Day Adventures*, Dole Food Company, 155 Bovet, Suite 476, San Mateo, CA 94402, or fax (415)570-5250. Put your request on school letterhead and tell how many CDs are needed.)

Productivity Programs

- *Kid Pix* or *Kid Pix Studio*
- *The Amazing Writing Machine*
- *HyperStudio*

Procedure:

Into: Before the Computer

- Using the transparency with the food group headings and referring to the food pyramid chart, have each child tell what he/she ate for breakfast. Have each child determine under which headings the foods belong and then write the foods on the transparency. After at least ten have responded, review the headings and foods under each.
- Referring to the food pyramid, discuss the number of servings suggested for each food group.
- Have students work in groups to plan a healthful menu for one day.

Integrated Lesson Plans

NUTRITION *(cont.)*

Into: Before the Computer *(cont.)*

- They may use the crayons to draw the various foods on the paper plates. Then allow each group to describe their menu to the class.
- Bring a community speaker (e.g., nutritionist, food service worker, chef, etc.) to talk to the children about good nutrition.
- Ask the children to plan their breakfasts and lunches for several days so that they begin to eat a more healthful diet.

Through: On the Computer

- Help the children use planning sheets (see Storyboards pages 142 and 143) before creating a slide show on the computer.
- Students work in groups to create slide shows on good nutrition.
- The children should include each of the food groups in their slide shows.
- Suggest that the children take turns recording the narration for each slide. Not only is this fun but it allows each student a turn.
- Encourage the students to choose appropriate clip art for each slide, e.g., an avocado for the fruit slide will work fine. The kitchen could work for cleaning vegetables or baking bread.
- As long as every food group is included in their presentations, this will indicate understanding of the food pyramid.
- They will need at least two class periods to complete their slide shows.
- Allow the opportunity to share the slide shows.

Beyond: Extra Activities

- Divide the class into the food groups.
- Each group will be in charge of bringing the food for that food group to class on the day of the "nutritious feast." They need to decide who brings what on the day of the feast.
- It will probably be wise not to assign a sweets group. Perhaps the teacher could bring in Popsicles for dessert.
- Use the assessment on the next page.

Integrated Lesson Plans

NUTRITION *(cont.)*
Assessment

Name _____ Date_____

                                                                                        Yes   No

I followed directions._____

We planned our work before using the computer._____

I did my part in our group. _____

We created a well-balanced menu. _____

Now I can help others with this project. _____

This is how I feel about the project. (Draw a face that shows how you feel.)

```



```

© Teacher Created Materials, Inc.   31   #2427 *Integrating Technology into the Science Curriculum*

Integrated Lesson Plans

NUTRITION *(cont.)*

Meat	
Dairy	
Fruits/Vegetables	
Grains	

Integrated Lesson Plans

LIVING THINGS

Whether introducing living things for the first time in kindergarten or as review in first through third, hands-on activities along with visual activities help clarify the terms living and non-living for primary students.

Materials:

- posters depicting a typical neighborhood or school setting
- drawing paper
- crayons

Living Things Reference Software
- *Plants CD*

Productivity Programs
- *Graph Club*
- *Microsoft Drawing*
- *Spreadsheet*
- *Kid Pix Studio*
- *Crossword Companion*

Procedure:

Into: Before the Computer

- While observing a poster of a typical neighborhood or school setting, have children point out various people, plants, animals, toys, vehicles, buildings, etc.
- Talk about which are living and non-living.
    - What makes something living?
        - -Living things can grow.
        - -Living things can move on their own.
        - -Living things can change.
    - What makes something non-living?
        - -Non-living things cannot grow.
        - -Non-living things cannot move on their own.
- After much discussion, pass out drawing paper and crayons and ask children to draw pictures which include living and non-living things.
- When finished with their drawings, the children can count how many living and non-living things are included in the drawings. They write the words with the numbers: Living—10, Non-Living—8

*For this example *Kid Pix Studio*, *Microsoft Works Spreadsheet*, and *Crossword Companion* were used.

Integrated Lesson Plans

LIVING THINGS (cont.)

Through: On the Computer

- Students use the drawing tool or multimedia software to duplicate their pictures on the computer.
- Name and save the pictures with a .bmp extension.
- Import the pictures onto a word processor page so that the children can write sentences explaining their counting of the objects. (below)
- Using graphing software or a spreadsheet, the children can graph their findings.
- Then they can create and print a chart which shows the number of their living and non-living things. (See the next page.)

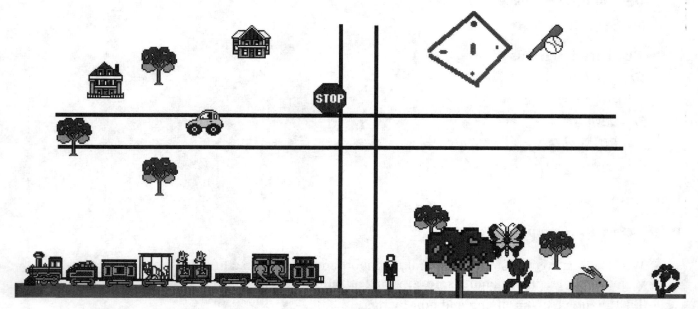

There are 17 living things in this picture. People, trees, flowers, grass, rabbit, elephants, giraffes, and tigers can grow or move on their own.

There are 17 non-living things in this picture. Trains, tracks, automobiles, stop signs, balls, bats, and houses cannot grow or move on their own.

Beyond: Extra Activities

- Begin a living/non-living bulletin board with pictures children cut out from magazines.
- They can work in groups and make collages of pictures about the following:

    Living Things in Our Homes

    Living Things in Our Neighborhoods

    Living Things in School

    Non-Living Things in Our Homes

    Non-Living Things in Our Neighborhoods

    Non-Living Things in School

- Have them use the draw tool and draw the pictures for the bulletin board.

LIVING THINGS (cont.)

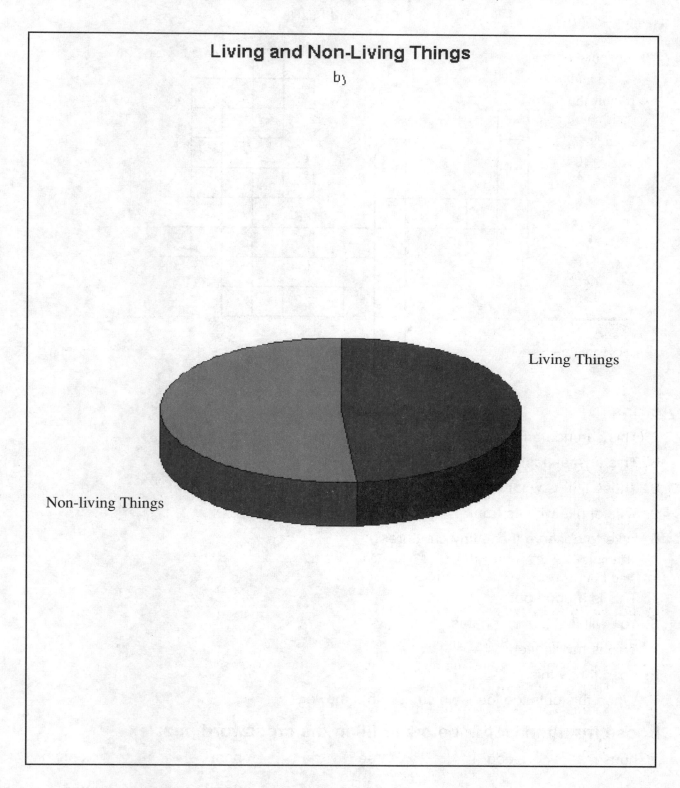

Integrated Lesson Plans

LIVING THINGS (cont.)
Crossword Puzzle

Across

2. This can ride on the water.
6. This animal has stripes.
8. These tell us what to do.
9. This drinks with its trunk.
10. I like to observe these tiny creatures.

Down

1. This is a good pet.
3. You will find leaves on this.
4. This is my shelter.
5. I can throw this.
7. When this gets too long, we cut it with a mower.

Choose from the words below to fill in the crossword puzzle:

| house | dog | tree | boat | elephant |
| baseball | grass | signs | tiger | bugs |

*Circle the things in the puzzle that are living.

Integrated Lesson Plans

PLANTS ARE LIVING THINGS

By "planting" stones and seeds side by side, students will be able to determine that of the two, plants are living things that grow. As the sunflower seeds sprout, the children can chart their growth, using observation, measurement, prediction, inference, and interpretation of data. As the children create a spreadsheets to chart their activities, they will have used all of these process skills with this enjoyable activity.

Materials:

- sunflower seeds
- small stones
- recycled foam produce trays or flower pots
- soil
- growth chart

Plant Reference Programs

- *Plants CD*
- *Sammy's Science House CD*

Productivity Programs

- *Microsoft Works Spreadsheet*
- *Kid Pix Studio*
- *The Graph Club*

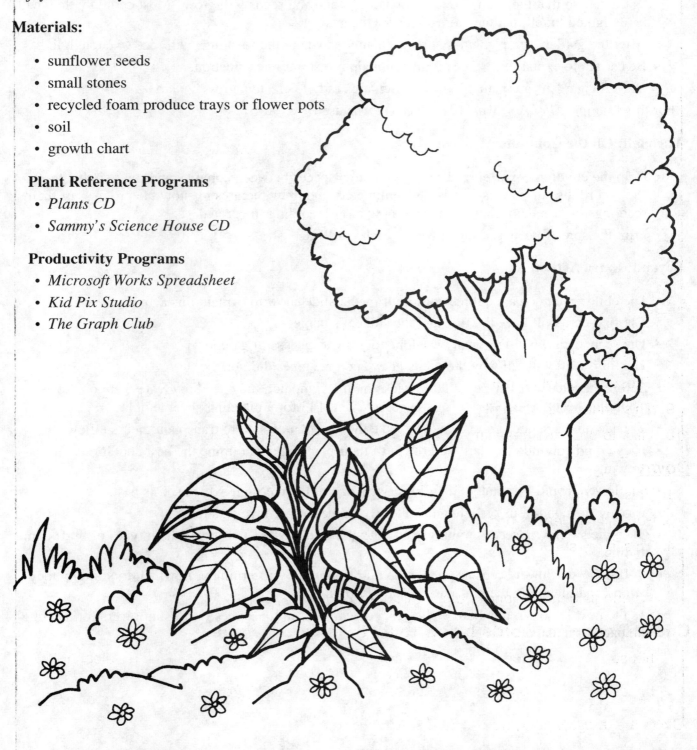

Integrated Lesson Plans

PLANTS ARE LIVING THINGS (cont.)

Procedure:

Into: Before the Computer

- Give each student a produce tray, stones, seeds, and soil. Instruct the children to plant their "gardens" so that they can determine whether stone and seeds will grow. Most certainly they will be delighted to tell that they already know the answer.
- After the gardens are planted, put them in an area where the seedlings will receive enough light.
- Be certain to explain that the soil must remain moist until germination.
- Once the gardens are planted, the children are ready to create a growth chart.

*For this example *Microsoft Works Spreadsheet* was used.

Through: On the Computer

- Help the children create spreadsheets with large spaces between columns and rows so that they can record the sunflowers' growth in centimeters and draw pictures to show the progress of their seedlings, or use the following page template for recording the growth.
- Print the spreadsheets and tape them to each garden.

Beyond: Extra Activities

- The children can create a three-card multimedia slide show to explain this activity.
- The first card will have the title and the student's name.
- The second card will tell what was done to grow a successful garden.
- The third card will tell why the seeds grew and the stones did not.
- Before students actually begin their stacks on the computer, they need to create them on the planning sheets. (See Storyboards pages 142 and 143 for a planning sheet template.)
- If space on the school grounds allows, take the children outside to transplant their sunflowers.
- They could continue to chart the growth of their plants and determine the best conditions for growth.
- Sunflowers could be planted in various types of soil to determine which soil is best.
- Varying the conditions of light and water would be further investigations.
- Once the sunflower seeds are harvested, install a bird feeder to observe which types of birds come to dine.
- Or use the sunflower seeds as part of the nutrition unit by determining in which food group they belong and including them as part of the nutritious feast.
- As the next lesson in this book indicates, you can certainly use these very sunflowers for labeling the parts of a plant.

PLANTS ARE LIVING THINGS (cont.)

Name: _____	My Sunflower Garden		
	Date	Height in cm	Drawing
Planted			
at 7 days			
at 14 days			
at 21 days			
at 28 days			

Draw a large picture below of your sunflower 21 days after you plant your seeds.

Integrated Lesson Plans

THE PARTS OF A PLANT

Children are more fearless than adults when it comes to experimenting on the computer. They will try and try again, and they are not discouraged as easily as adults are when things don't work out just right. Allowing them to draw plants on the computer is fun. Be sure that their plants are what they want. Do not worry about the colors, sizes, or dimensions—their plants can even be imaginary ones. In that case, their plants will surely have all of the right "parts." Even while experimenting, they will draw on their experiences with the plants that they have seen.

Materials:

- many posters and pictures of plants
- books and magazines with pictures that the children can take to their tables/desks
- videos of plants—flowers, trees, vegetables of all kinds
- drawing paper and crayons
- assessment (See the sample at the end of this lesson.)

Productivity Programs
- *Paint*
- *HyperStudio*
- *Kid Pix Studio*

Procedure:

Into: Before the Computer

Allow the children lots of time to peruse the pictures of plants.
- Ask questions:

    Which are your favorites?

    Where are they located?

    What parts do they have in common?

- If there are flowers and trees on the school grounds, take a walk outside and look at as many plants as possible.
- Bring in some plants from home if a walk outside is not possible.
- Provide time for students to draw plants, real or otherwise. (See the next page for a sample.)
- This might take an entire class period.

#2427 Integrating Technology into the Science Curriculum   40   © Teacher Created Materials, Inc.

THE PARTS OF A PLANT (cont.)
Sample Drawing

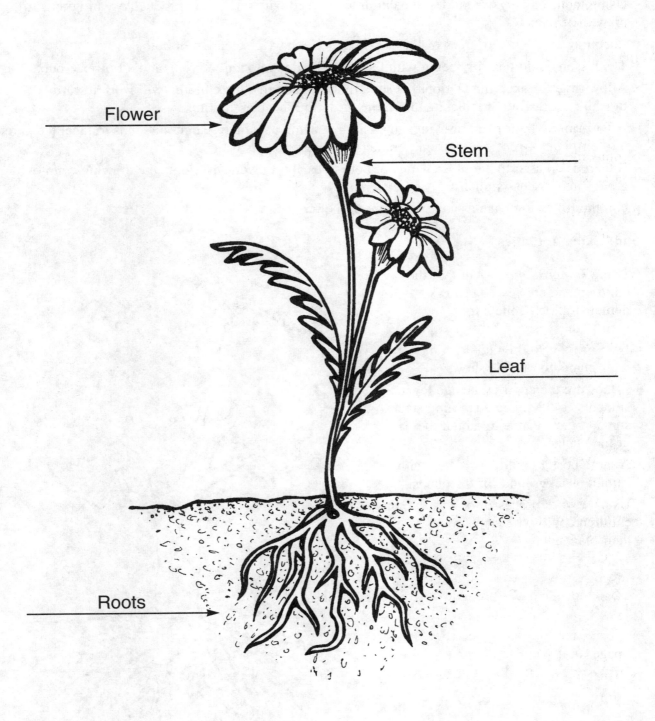

Integrated Lesson Plans

THE PARTS OF A PLANT *(cont.)*

*For this example *Paint* was used.

Through: On the Computer

- Using their drawings as guides, the children proceed with one of the productivity programs to draw their plants.
- Encourage drawings of trees or flowers so that you have a selection of both.
- The lesson could possibly begin with the class divided into a tree group and a flower group.
- Allow time for experimentation, but set a time limit so that the children will work toward finishing a drawing of a plant and printing it during this class period.
- After printing their plant drawings, the children again use the research materials to label the parts of their plants—roots, stem/trunk, branches, leaves, flowers/fruits.
- If there is an assortment of drawings of both trees and flowers, assemble the drawings into two groups and discuss similarities and differences.
- Use the following page as an assessment of the project.

Beyond: Extra Activities

- Using *HyperStudio* or *Kid Pix*, the children can create slide shows demonstrating habitats for their plants.
- Four or five children could combine their plants for a slide show.
- If they are the appropriate age level, students could create a learning stack so that others can learn about the parts of the plants.
- A large poster could be created with a suitable background for the plants.
- Using a word processor or *Kid Pix*, the children could type labels for their plants.

Integrated Lesson Plans

THE PARTS OF A PLANT *(cont.)*

Use the following words to label the parts of the tree and the flower. You may use a word more than once:

Trunk   **Stem**   **Branch**

Leaf   **Roots**   **Flower**

The Parts of a Flower

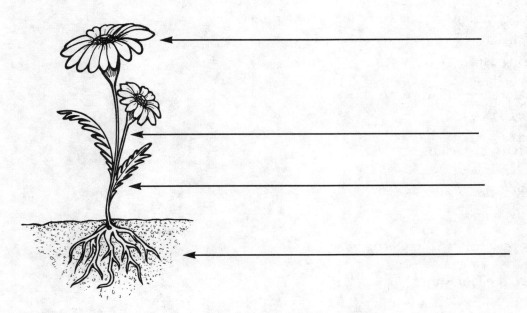

The Parts of a Tree

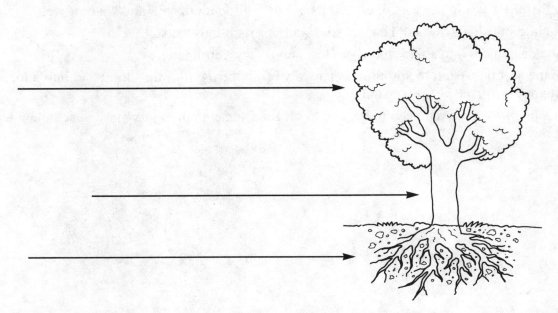

© *Teacher Created Materials, Inc.*   43   #2427 *Integrating Technology into the Science Curriculum*

Integrated Lesson Plans

WHAT DO PLANTS NEED TO GROW?

Once the children begin to plant and care for their own plants, they will begin to create hypotheses about the process. This activity uses controls to better help them form conclusions about how plants grow. If appropriate for the age group, they will then create multimedia slide shows telling what plants need to grow.

Materials:

- many books about plants
- seeds
- soil
- foam egg cartons or small containers to start seeds
- craft sticks
- watering can

Plant Reference Software
- Plants CD

Productivity Programs
- *Paint*
- *HyperStudio*
- *Kid Pix Studio*

Procedure:

Into: Before the Computer

- If students have not yet had the joy of planting, now is the time. Give each child a container, at least three seeds, some soil, and a craft stick.
- Help them put soil in the container, add the seeds, a little more soil, and water it well.
- Each child puts his name on the craft stick and inserts it into the soil.
- Put the containers into a dark room or closet until they germinate.
- Once the seedlings start to sprout, immediately bring them out of the closet and into a location where they will receive ample sunlight.
- At this point the children need to be sure to give the same amount of water to each plant every three days.

WHAT DO PLANTS NEED TO GROW? *(cont.)*

Into: Before the Computer *(cont.)*

- There should be four extra containers for the controls:
    - One has no soil but receives same amount water and light as the rest.
    - One has soil and light but no water.
    - One has soil and water but is placed in a dark room or closet for the duration of the activity. (This one will germinate, but then it will wither and die.)
    - One has soil and water until germination; when taken out of the closet, it no longer receives any more water.

*For this example *HyperStudio* was used.

Through: On the Computer

- The children carefully care for their plants for about three weeks.
- After the children observe all of the plants, including the controls, generate some discussion as to what plants need to grow.
- The children will work in small groups to create a six-card stack telling what plants need to grow. (See page 9 regarding delegating tasks within groups.)
- The first card is the title card and has a bright, yellow sun drawn on it with the paintbrush tool. The title uses a text box which says, "Plants need light to grow," and the background is a light blue.
- The first card is the only one that has a button to click. It will play CNTRYSTP.WAV; this is a little song that will play throughout the entire six cards.
- The second card will have a mound of grass painted on it with a sun painted in a different position. (See the next page for a screen shot of all six cards.) The beginning of a stem can be painted in, as well as some brown soil around the stem. This card can then be copied four more times. The text box will say, ". . . and soil."
- The second through fifth cards will have invisible buttons and automatic timers to go to the next card after three seconds.
- Clouds and wind lines can be added to the third card with the words, ". . . and air."
- Add some darker blue spray can splotches for rain on the fourth card which says, ". . . and water."
- Just paint some leaves on the fifth card.
- And the sixth card finally gets the bright red, orange, or purple flower. This card's button will have applause and not go to any card.
- Naturally, there are many choices of buttons, paint colors, etc., when the children create their stacks. The above directions are merely suggestions.
- The children will probably want to share their stacks with parents and other classes. Sometimes the best part of creating is sharing with others.

Integrated Lesson Plans

WHAT DO PLANTS NEED TO GROW? *(cont.)*

Beyond: Extra Activities

- As in the previous lesson, Plants Are Living Things, the children might want to chart the growth of their plants.
- Again, if appropriate, plants can be planted on the school grounds.
- By integrating writing, the children could use the word processor to write "how to" paragraphs on what plants need to grow.

HyperStudio **Stack**

(*HyperStudio*, Roger Wagner Publishing)

Integrated Lesson Plans

ANIMALS ARE LIVING THINGS

There is no better way to explore animals as "living things" in the primary classroom than to actually have a classroom experience with the birth of baby animals. If a class can experience this in any way, it should. Chrysalides transforming into butterflies, tadpoles suddenly appearing in pond water, the birth of baby gerbils, peeping chicks in an incubator—all of these are wonderful experiences for primary age children. This lesson will demonstrate using butterflies since they are one of the easiest animals to observe. They also take very little class time.

Materials:

- reference books and library books about butterflies
- pictures and posters showing various butterflies
- Butterfly House (Call 1-800-livebugs for information on ordering.)
- drawing paper and crayons or markers
- larger poster or bulletin board paper

Reference Programs
- *The Animals! 2.0 CD*
- *Zurk's Rainforest Lab*

Productivity Programs
- *Kid Pix Studio*
- *Paint*

Procedure:

Into: Before the Computer

- The Butterfly House includes directions on how to send for chrysalides. When they arrive, everything you need to successfully hatch your butterflies has been included.
- The time span from the arrival and setup to the release of the butterflies is approximately three weeks.
- Have your students draw pictures at each stage:
    - Setting up the house
    - Putting the caterpillars in the vials with their food
    - Taking the chrysalides out of the vials and laying them on the lids in the house
    - Hatching inside the house
    - Placing flowers and food in the house when the first butterflies hatch
    - Releasing the butterflies on the school grounds
- Depending on grade level, the children can write words, sentences, or paragraphs to explain their pictures.

Integrated Lesson Plans

ANIMALS ARE LIVING THINGS (cont.)

*For this example *Kid Pix Studio* was used.

Through: On the Computer

- Use the storyboard templates provided on pages 142 and 143 before beginning work on the computer.
- Using *Kid Pix Studio*, the children can draw their interpretations of the butterfly house. Then they can print them out and tape them onto posters or thumbtack them onto the bulletin board.
- The children can continue the poster until the butterflies are released.
- It is possible, group the children so that each child in the group draws one picture for the group. The outcome would be five different posters completed for five groups, and each would be a little different.
- If desired/appropriate, use the assessment on the next page for this project.

Beyond: Extra Activities

- Once released, many of the butterflies will remain in the area if the food they require is available.
- Research what flowers butterflies like, both for food and for laying eggs. Start a butterfly garden with donations of plants from parents. When they are dividing perennials, many people would be happy to share.
- Determine a "before and after"—what animals visited your area before the planting of special plants and then after the planting.
- Start a butterfly watchers club. Use a butterfly book to identify the various kinds of butterflies spotted and chart them.

Internet Connections:

- http://4Kids.org/coolspots/creatures/
- mgfx.com/butterfly

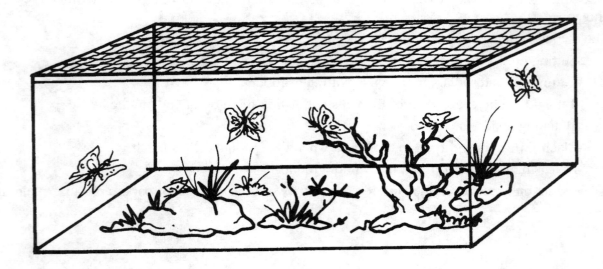

ANIMALS ARE LIVING THINGS (cont.)

Assessment

Name _____ Date _____

Directions: Circle the answer that best fits.

1. I enjoyed this project about butterflies.

    YES   NO

2. I know animals are living things because

    they have beautiful colors.

    they move on their own.

3. The best part of this project was

    setting it up.

    watching the butterflies.

    drawing the pictures on the computer.

4. I would like to do something like this again.

    YES   NO

5. I can tell two things about butterflies after this project:

    a. _____

    _____

    _____

    b. _____

    _____

    _____

Integrated Lesson Plans

ORDERING ANIMALS

Imagine this picture—a flea standing next to an elephant. This activity will evoke smiles from the children as they draw pictures of large and small animals on the computer. It will work well as a cooperative learning activity since they will help each other name, draw, and then order to size as many animals as will fit on the computer screen.

Materials:

- reference books and library books about animals
- pictures and posters showing various animals

Animal Reference Programs
- *The Animals! 2.0 CD*
- *Zurk's Rainforest Lab*
- *Oceans Below CD*

Productivity Programs
- *Kid Pix Studio*
- *The Graph Club*
- *Paint*

Procedure:

Into: Before the Computer

- Group the children with about five or six in each group. (See grouping and tasks on page 9.)
- Allow the children time to explore the research materials so that they can list many animals that they will want to draw on the computer.
- Each group should have a list of about four or five animals. They can list some of the same animals in each group.
- Each person in the group should draw some of the animals.
- Before going to the computer, students need to order their animals in some way: smallest to largest or largest to smallest.
- They need to have a point of reference for graphing, e.g., one hand high, one finger high, one eraser high, etc.

*For this example *Kid Pix Studio* and *Paint* were used.

ORDERING ANIMALS (cont.)

Through: On the Computer

- Each child can choose to use either the stamps or the drawing tools.
- When using stamps, they need to understand that the stamps might all be the same size, but the animals pictured on the stamps are very different in size. Keep the poster and pictures handy for reference.
- Each group will create the pictures on one screen, being sure that the animals are lined up in the correct order.
- When they have their pictures ordered, they will, if appropriate, write sentences explaining the ordering.
- They will print the pictures.

The dog is larger than the frog, and the horse is larger than the dog.

The robin is smaller than the osprey, and the osprey is smaller than the eagle.

- Using graphing software or a spreadsheet, they will use their measuring reference to graph their animals:
    - one picture on the graph equals the size of a hand, or
    - one box equals the size of a hand.
- See the sample of a graph on the next page.

Integrated Lesson Plans

ORDERING ANIMALS (cont.)

Through: On the Computer *(cont.)*

A frog is one hand. A dog is five hands. A horse is 15 hands.

Frog	Dog	Horse

Beyond: Extra Activities

- The children can use graphing software to show a number of animal parts.
    Legs: snake, 0; bird, 2; alligator, 4
- The number of young can also be graphed.
- Graph the numbers of various animals seen in the course of a week. This will help the children become more observant of their surroundings.

ANIMAL GROUPS

Use a jigsaw method of collaborative learning to make the best use of time. The number of groups is determined by the number of topics to be learned. Each student in a group becomes the expert on that topic and presents to the other groups. Having the group create the assessment (when appropriate) for those it teaches ensures that all areas of the topic are learned. Animal classification is one of many topics where jigsaws work well.

Materials:

- Reference books and library books about animals and animal classification
- Pictures and posters showing various animal classification
- Magazines containing animal pictures that can be cut out (Let parents know in the beginning of the year that you will accept all old *National Geographics*, etc., for this purpose.)
- Larger poster or bulletin board paper

Animal Reference Programs

- *Animals CD* (Clearvue)
- *Animals! 2.0* (Mindscape)
- *Zurk's Rainforest Lab*
- *Animal Classifications* (Centre Communications)

Productivity Programs

- *Kid Pix Studio*
- *Paint*
- *The Amazing Writing Machine*
- clip art programs

Procedure:

Into: Before the Computer

- Put the children in mixed ability groups for this activity. Explain that they will become the experts about a particular animal group. Four or five children would be the maximum number in each group.
- Assign each group an animal classification: birds, mammals, insects, reptiles, amphibians, fish. Depending on class size, arachnids may be included as a special group. (Children love to study spiders!)
- Each child in the group has a special topic to research about the group's animals:
    - Habitat
    - Description
    - Birth and care of the young
    - Interaction with like and unlike animals

Integrated Lesson Plans

ANIMAL GROUPS (cont.)

Into: Before the Computer (cont.)

- One or two children in the group can be responsible for making sure all topics are being covered, the materials are available and cared for, finding the pictures or graphics, and putting together the final presentation, etc.
- Use the planning worksheets on page 55 for the students' research.

*For this activity *The Amazing Writing Machine* was used.

Through: On the Computer

- Each child chooses a graphic or picture from a magazine to include with the topic researched.
- Print out the paragraphs.

Beyond: Extra Activities

- Four or five copies of each topic could be printed, collated, and stapled together so that each member of the group has a book about his animal classification.
- Each person makes a cover for his book.
- Each group presents its animal classification to the class.
- Put children into groups so that each animal classification is represented in every group.
- The jigsaw portion comes together so that an expert on each classification is working in every group for further discussion.

Which Animal Group Is Missing?

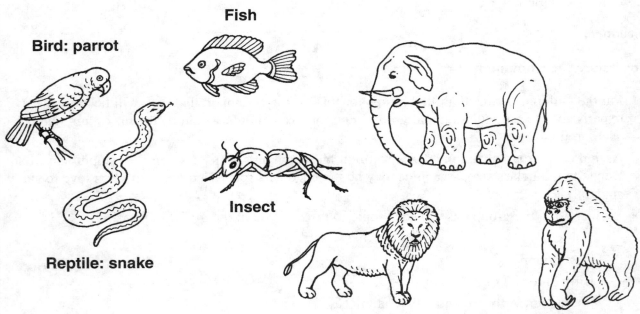

Answer: amphibian

Integrated Lesson Plans

ANIMAL GROUPS (cont.)

Animal Classification Planning Sheets

clip art or picture from a magazine

Animal:_____ Classification:_____

Topic:_____

_____
_____
_____
_____
_____
_____
_____
_____
_____
_____
_____
_____
_____
_____
_____
_____
_____
_____

Integrated Lesson Plans

ANIMAL GROUPS (cont.)

The Birth and Care of Young Amphibians

Animal: Salamander   Classification: Amphibian

Topic: Birth and Care of Young

The salamander lays eggs in water. The eggs are soft and look like jelly. After the salamander lays the eggs, she swims away. When the eggs hatch, they are on their own. The mother and father do not care for the babies. Some eggs do not hatch. Do you know why? They are eaten by fish and other animals. For this reason the mother salamander lays many eggs.

Integrated Lesson Plans

ANIMAL GROUPS (cont.)

What Do You Know About Animals?

Name _____ Date _____

Directions: Circle all of the correct answers. There might be more than one for some questions. Then give one reason why.

1. Which of these animals is a mammal?
   robin   snake   rabbit   ant

Why?_____

2. Which of these animals is a bird?
   owl   turtle   bat   shark

Why?_____

3. Which of these animals is an amphibian?
   blue jay   cow   frog   snake

Why?_____

4. Which of these animals is a fish?
   monkey   shark   crow   whale

Why?_____

5. Which of these animals is a reptile?
   turtle   hummingbird   iguana   tadpole

Why?_____

6. Which of these animals is an insect?
   bee   spider   ant   roach

Why?_____

© Teacher Created Materials, Inc.

Integrated Lesson Plans

ANIMAL GROUPS (cont.)

How Did I Do?

Name_____ Date_____

	Yes	Sometimes	No
I worked hard to research my topic.	___	___	___
I cooperated with my group.	___	___	___
I used my time wisely.	___	___	___
I did a good job presenting my topic.	___	___	___
I listened and learned when others presented.	___	___	___
I can classify animals.	___	___	___

The best part of this activity was _____.

The part I liked least was _____.

The next time I have to do research, I will_____
_____

The animal I like best from this activity is the _____because
_____
_____

Integrated Lesson Plans

THE PARTS OF AN INSECT

Just as they can identify the parts of a plant, primary students enjoy identifying parts of animals. Teaching to all learning styles continues to be critical. In this lesson the children will capture and examine various insects, discuss them, read about them, write about them, and, using a draw software, print out their scientific and artistic interpretations of them. Learning about the importance of insects to our environment will also help them understand why they should treat them with respect and release them when through with their research.

Materials:

- reference books and library books about insects
- pictures and posters showing various insects
- live insects brought in by students or live insects caught on school grounds
- drawing paper and markers/crayons

Insect Reference Software

- *The Animals! 2.0 CD*
- *Zurk's Rainforest Lab*
- *Junior Nature Guides: Insects* (ICE)

Productivity Software

- *Kid Pix Studio*
- *Draw* or *Paint*

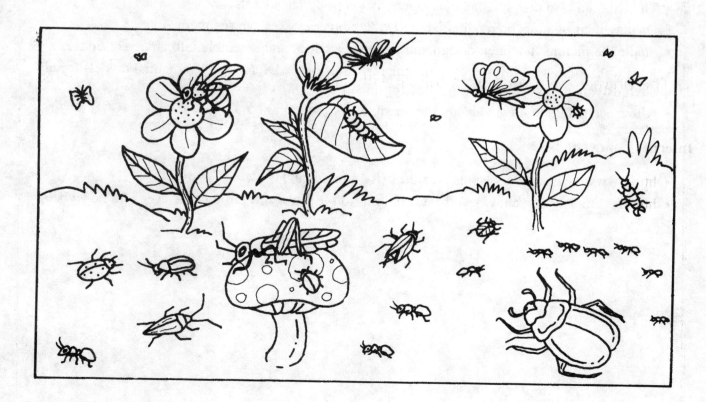

Integrated Lesson Plans

THE PARTS OF AN INSECT (cont.)

Procedure:

Into: Before the Computer

- Tell the children they are going to study live insects, and they need to bring one from home. They need to put it in a jar with one little hole punched in the lid.
- Or, you can take them out on the school grounds with jars and nets if you'd like. They should not attempt to bring in bees or wasps, since they can sting.
- Tell them they should not bring in spiders because they are not insects; they may already know this if the class has previously studied animal classification.
- Using the research materials, they can attempt to identify their insects.
- Have each child draw his/her interpretation of his/her insect, labeling its parts.

*For this activity *Paint* was used.

Through: On the Computer

- The children draw their interpretations of their insects on the computer, label the parts, and print their drawings.
- If appropriate, they can write sentences or paragraphs explaining the uses of the insect's parts.

Beyond: Extra Activities

- The children also draw the habitats where they found their insects.
- Display their pictures on a bulletin board where other classes can see them.
- Staple the pictures together, design and laminate a cover, and assemble into an *Insect Book*.
- Repeat this activity with the other animal classifications. (Do not use live animal models for all; obviously, pictures will have to suffice for most.)
- Allow each child to create his/her own book of *Animal Parts*.

Internet Connections:

- http://www.4Kids.org/coolspots/creatures (for all animals)
- http://cvs.anu.edu.au/andy/beye/beyehoe.html (to see the world through the eyes of a honeybee)

Integrated Lesson Plans

THE PARTS OF AN INSECT *(cont.)*
Parts of An Insect

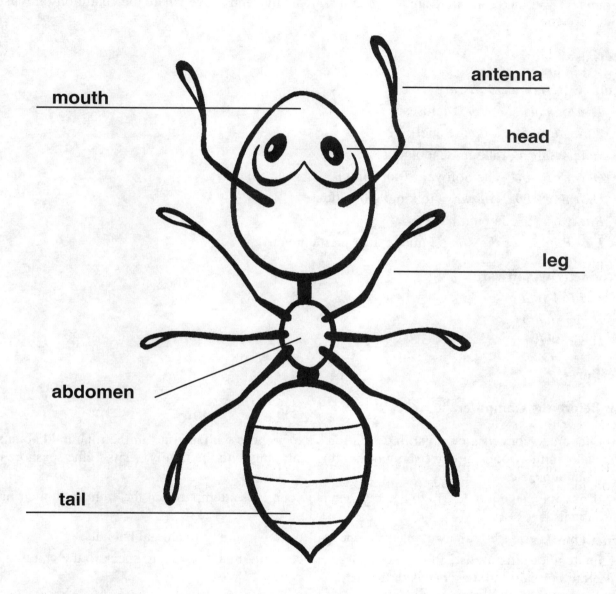

An insect is divided into three parts.

Its head is one part, its abdomen is one part, and its tail is one part.

An insect has two antennae, two eyes, and one mouth.

It has six legs.

Integrated Lesson Plans

ANIMAL HABITATS

Most children have experienced studying animals in their own backyards or at the zoo or on nature programs on TV; therefore, they are aware that animals live and thrive where the conditions are just right.

Materials:

Animal Reference Software
- *The Animals! 2.0 CD* (Mindscape)
- *Zurk's Rainforest Lab* (Soliel)
- *Jungle Safari* (Talking Schoolhouse)
- *Oceans Below* (The Software Toolworks)
- *Learning About Animals* (Talking Schoolhouse)
- *Backyard* (Brøderbund)
- *Putt-Putt Saves the Zoo* (Humongous Entertainment)

Productivity Software
- *Kid Pix Studio*
- *HyperStudio*
- *Draw* or *Paint*

Procedure:

Into: Before the Computer

- Encourage conversation regarding children's experiences with animals in their natural habitats: walks in the woods, going fishing or hunting, camping with the family, visits to the seashore or mountains, etc.
- If possible, take a field trip to a zoo to see how our zoos attempt to put the animals in their natural habitats.
- Allow experience with available reference software pertaining to animal habitats.
- Invite a local naturalist, zoo worker, park service representative, etc., to speak to the children about animals and their needs in the wild.
- Information about introducing the red wolf back into eastern North Carolina is a valuable topic with videos and lesson plans available from *North Carolina Wildlife Notebook*.
- Give the children copies of the blackline master on page 65 so that they can take notes about the various animal habitats as you research together. They will draw or write sentences, depending upon their abilities.
- The children decide on the habitat and animal and then sketch their first draft.

Integrated Lesson Plans

ANIMAL HABITATS *(cont.)*

*Kid Pix Studio was used for this activity.

Through: On the Computer

- Use the storyboard planning sheets before beginning to create slides for the habitat slide show.
- The children create habitats for their animals, using their notes and reference materials as needed.
- They can use stamps or clip art, if available, to add their animals to their habitats.
- They can draw their animals if the desired stamp or clip art is not available.
- Save each student's habitat as a slide for a habitat slide show.
- Either the teacher or the students create a title page for the slide show.

A *Kid Pix Studio* Slide Show

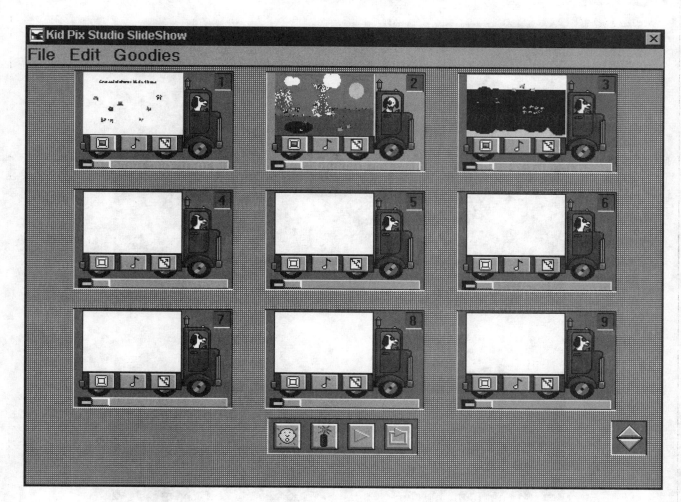

© *Teacher Created Materials, Inc.*   63   *#2427 Integrating Technology into the Science Curriculum*

Integrated Lesson Plans

ANIMAL HABITATS (cont.)
Habitat Slide Show
Animal Habitat Slide Show

Integrated Lesson Plans

ANIMAL HABITATS (cont.)

Notes for My Animal Habitat

Name _____ Date _____

I have chosen to research the habitat of _____

Where in the world is my habitat?
_____

The climate is_____
_____

The plant life located there is_____
_____
_____
_____

The other animals I might find in this habitat are _____
_____

My animal likes this habitat because _____
_____
_____
_____

© Teacher Created Materials, Inc. 65 #2427 *Integrating Technology into the Science Curriculum*

Integrated Lesson Plans

FOOD WEBS

Creating food webs and food chains helps primary age children understand the interdependence of organisms in our environment. Drawing food webs with the help of technology makes learning about them fun.

Materials:

- reference books and science texts with pictures depicting food chains and food webs
- posters showing food chains and food webs
- drawing paper and crayons/markers

Interdependent Reference Software

- *Plants CD* (Clearvue)
- *A Field Trip to the Rainforest* (Sunburst)
- *Backyard* (Broderbund)

Productivity Software

- *Kid Pix Studio*
- *HyperStudio*
- *Draw* or *Paint*

Procedure:

Into: Before the Computer

- Allow time for research and discussion of what the animals in the various food groups eat.
- Discuss producers, consumers, predators, prey, scavengers, etc.
- Using the posters and research texts as reference, allow the children to draw pictures of various biomes (habitats) and the plants and animals that live there. Have them share their drawings in groups, discussing the producers and consumers present in their drawings.
- Then each child will choose the food web he/she wants to illustrate on the computer after using the planning sheet on page 67.

*This activity used *Kid Pix Studio*.

Through: On the Computer

- Using their drawings as guides, the children can now create food web illustrations with one of the productivity software programs.
- If appropriate they can label producers, consumers, etc., or write sentences for their illustrations.

FOOD WEBS (cont.)

Food Web Planning Sheet

My habitat is _____

Some producers or plants that I can include in this habitat are_____
_____

Some animals that live here are_____
_____

I choose _____ for my producer. I will choose _____

and _____ for my prey and predators.

Draw a picture of your food web below:

FOOD WEBS (cont.)

A Meadow Food Web

The grass and plants are producers. They make their own food with help from the sun's energy. The rabbit, insect, and fish eat the grass. They are consumers. The fish is also a predator since it eats the insects.

The rabbit and frog are prey for the snake, who is a predator.

The eagle is a predator, too. It eats the snake and the rabbit.

A food web shows the interdependence of living things.

Integrated Lesson Plans

PREDATOR/PREY FANTASY

The children will use their knowledge about predators and prey in their natural habitats to create fantasy plants and animals and put them in a habitat that they imagine for their pretend plants and animals. Writing across the curriculum is encouraged, and the science curriculum lends itself to wonderful fantasies created from the minds of primary children. This creative writing activity could be one type of assessment for the topic.

Materials:

Reference Software
- *Plants CD* (Clearvue)
- *Zurk's Learning Safari* (Soleil)
- *A Field Trip to the Rainforest* (Sunburst)
- *Backyard* (Broderbund)
- *Type Right* (Type Right)
- *Stickybear Typing* (Optimum Resource)

Productivity Software
- *Kid Pix Studio*
- *Microsoft Works*
- *The Amazing Writing Machine*

Procedure:

Into: Before the Computer

- It's never too soon to begin keyboarding. Children love to write their stories directly onto the computer even before they master the keyboard. Revision and proofreading are so much easier on the word processor than having to rewrite by hand.
- Remind children of some stories about animals and habitats, rereading any that you'd like to get them in the mood:

    The Very Hungry Caterpillar by Eric Carle

    Where the Wild Things Are by Maurice Sendak

    The Mouse and the Motorcycle by Beverly Cleary

    Frog and Toad books by Arnold Lobel

    The Velveteen Rabbit by Margery Williams

- Begin brainstorming for a fantasy about predators and prey by listing on the board as many animal characters as the children can mention.
- Brainstorm habitats where the story takes place.
- Then brainstorm action that could take place in the fantasy.

© Teacher Created Materials, Inc.

Integrated Lesson Plans

PREDATOR/PREY FANTASY (cont.)

The Amazing Writing Machine was used for this activity.

Through: On the Computer

- Have the children use the story web below before beginning their fantasies.
- After you have checked their webs, allow them to get right on the computer to begin their stories.
- They can use clip art or *Paint* software for their pictures for their stories.
- Or they can use a storywriter like *The Amazing Writing Machine*. This can be folded into a book like the example on pages 71 and 72.

Beyond: Extra Activities

- Put the children's stories together into a book of *Predator/Prey Fantasies*.
- Have the children use their stories to make a multimedia presentation in which they put the pictures on the cards and record their stories.
- Have them act out their stories in groups.

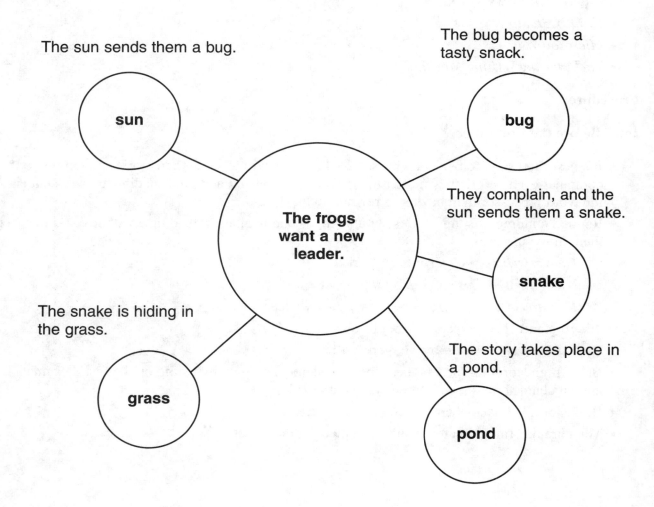

PREDATOR/PREY FANTASY *(cont.)*

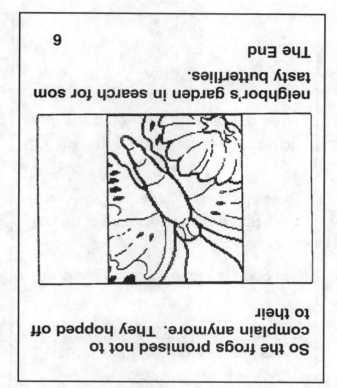

6

So the frogs promised not to complain anymore. They hopped off to their neighbor's garden in search of some tasty butterflies.
The End

1

The Complaining Frogs

by:

Integrated Lesson Plans

PREDATOR/PREY FANTASY (cont.)

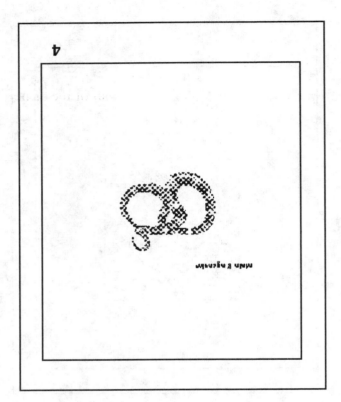

3

But the Sun said, "Snakes are part of your world now. So you have to be careful, or it will eat you, just as the snake needs to watch out for large birds."

The frogs promised the Sun that if the snake kept away, they would stop complaining.

The snakes swam out of the pond to look for a tasty snack.

Suddenly, the snake moved and the frogs hid themselves in the pond grasses.

The frogs were tired of the same old food in their pond. So they complained to the Sun, and demanded something more interesting for them.

The Sun saw what silly frogs they were, so he threw down a huge snake into the pond. The frogs thought it was a log, and started to hop over to it to look for some tasty insects that might be crawling on the log.

2

#2427 Integrating Technology into the Science Curriculum   72   © Teacher Created Materials, Inc.

THE PLANT CYCLE

Rather than group this activity with the Plants section of this book, this lesson and the lesson on animal cycles appear with the lessons having to do with cycles and patterns. However, it may be used while teaching your students about plants. So much in our daily life is regulated by cycles: day and night, the seasons and weather patterns, water. They are all connected, and primary age children recognize these patterns and cycles. Perhaps it is the security of knowing what to expect that encourages them to readily learn about these cycles. In this activity the children will keep a computer journal of the "Life Cycle of a Bean."

Materials:

- research books, texts, poster, and pictures of mature, fruiting plants and their cycles
- drawing paper and crayons/markers
- fresh beans in their pods
- pebbles, soil, bean seeds, plant food, aluminum pie plate, plastic bag with tie
- large planting pots or an outside garden plot

Reference Software
- *Plants CD* (Clearvue)
- *Sammy's Science House* (Edmark)
- *A Field Trip to the Rainforest* (Sunburst)
- *Backyard* (Broderbund)

Productivity Software
- *Kid Pix Studio*
- *Draw* or *Paint*

Procedure:

Into: Before the Computer

- As a whole-class activity, have a student put the pebbles in the bottom of the pie pan.
- Another student puts most of the soil on top of the pebbles.
- Another lays the seeds on top of the soil.
- Another sprinkles the remaining soil over the seeds.
- Water well and slip the pie pan into a plastic bag.
- Have each student begin a "Life Cycle of a Bean" journal with a sketch of the pie pan, date, and, if appropriate, a sentence describing the planting of the seed.
- Put the bag in a closet until germination.
- Check the pan in about 8 to 10 days; if seedlings appear, bring it out of the closet. Take it out of the bag and put it near a sunny window.

Integrated Lesson Plans

THE PLANT CYCLE (cont.)

Into: Before the Computer *(cont.)*

- Allow the children to sketch the progress of the plants in their journals approximately every three days.
- When the seedlings have their "real" leaves (the first two leaves don't count), transplant them into individual containers. If space is available, begin a garden patch on the school grounds.
- Continue giving water and plant food as needed, continuing to sketch the progress of the plants.
- At any point along the activity, the children can begin their computer journals.

*Kid Pix Studio was used for this activity.

Through: On the Computer

- At this point the teacher may decide if the children should work in cooperative groups on the computer or individually, depending on the computer situation. We will assume that they are working in groups.
- With five students in each group, each student will complete one drawing of the plant cycle on the computer:
    * the bean in its pod and the pie plate with planted seeds
    * the pie pan near a window with only the first leaves of the bean plants
    * the transplanted bean plants with their "real" leaves
    * the bean plants after they have grown tall
    * the bean plants with bean pods on them
- Each child can create a computer-generated cover for his/her journal of "The Life Cycle of a Bean."
- Print out five copies of each journal and use brads, a plastic spiral binding machine, or staples so that each child has a copy of the group's journal.

Beyond: Extra Activities

- Create a slide show with the pages of the journal.
- Instead of publishing a book, tack the pages onto a "Life Cycle" bulletin board, showing both plant and animal (see next lesson) life cycles.
- Create a three-dimensional life cycle table. Have the children cut out their computer-generated drawings and paste them onto heavy construction paper. Bend the bottom under and paste or tape onto large poster board which is divided into life cycle boxes.

Internet Connections

http://www.vg.com (for lots of gardening information; however, more for teacher than students)

THE PLANT CYCLE (cont.)

Cover Page of Student Journal

My Journal of

The Life Cycle of a Bean

by

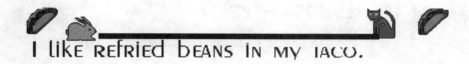

I like refried beans in my taco.

Page One of the Group's Journal

The Life Cycle of a Bean

DATE: _____

 A BEAN POD

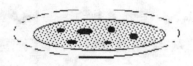

FIRST, WE PLANT THE BEAN SEEDS IN A PIE PAN. PUT THE PAN IN A PLASTIC BAG, AND PUT IT IN A CLOSET.

Integrated Lesson Plans

THE PLANT CYCLE (cont.)

Second Page of Group's Journal

The Life Cycle of a Bean

When the seeds form their first leaves, we take them out of the closet & put them near a window.

Now, they need sunlight, water, and plant food.
Date_____

Third Page of Group's Journal

The Life Cycle of a Bean

Date:_____

When our bean plants have "real" leaves, we transplant them outside in our garden plot. We are sure to give them food and water.

Integrated Lesson Plans

THE PLANT CYCLE *(cont.)*

Fourth Page of Group's Journal

The Life Cycle of a Bean

DATE:_____

ONLY ONE OF OUR PLANTS IS NOT HEALTHY AND GROWING.

Fifth Page of Group's Journal

The Life Cycle of a Bean

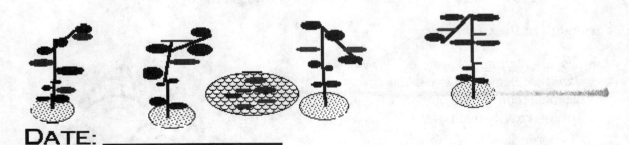

DATE: _____

WE HAVE PICKED BEAN PODS FROM OUR BEAN PLANTS. THE CYCLE IS COMPLETE.

THE LIFE CYCLE OF A MEALWORM BEETLE

Materials:

- recycled plastic containers with lids
- mealworm beetles and larvae (bought at pet or bait shops)
- bran or oatmeal, apple or potato

Reference Software
- *Multimedia Bug Book* (Expert)
- *Sammy's Science House* (Edmark)
- *Backyard* (Brøderbund)

Productivity Software
- *The Graph Club*
- *Microsoft Works Spreadsheet*
- *Kid Pix Studio*

Procedure:

Into: Before the Computer

- Put some bran or oatmeal cereal in the bottom of a recycled plastic container.
- Add a piece of apple or potato for moisture.
- Count the adult beetles and larvae as you put them into the container.
- Punch nail holes in the lid, and put the container in a safe place out of direct sunlight.
- Check the container every five days for about four weeks.
- Be sure to replace the apple or potato if it begins to get moldy.

Through: On the Computer

- Using graphing software or *Draw* or *Paint* software, allow the students to keep a record of the number of mealworms counted every five days.
- If appropriate, they can write sentences or a paragraph to explain the process.

"Where can I find a juicy beetle?"

Integrated Lesson Plans

THE LIFE CYCLE OF A MEALWORM BEETLE (cont.)

Beyond: Extra Activities

- Change the conditions of the mealworm colony and graph the differences:
    * Put one colony near a sunny window.
    * Put one in a dark closet.
    * Use a different cereal in one.

DAY 5

WE HAVE EIGHT BEETLES AND FIVE LARVAE.

DAY 10

WE HAVE TEN BEETLES AND SIX LARVAE.

DAY 15

WE HAVE 13 BEETLES AND NINE LARVAE.

DAY 20

WE HAVE 11 BEETLES AND FIVE LARVAE. WE LEFT MOLDY APPLE IN THE BEETLE HOUSE.

Integrated Lesson Plans

DAY AND NIGHT SKY

The different sights and sounds of day and night are familiar to young children. Observing and listing what they see and hear is the beginning of addressing this cycle. Moving on to why we have day and night and the earth spinning on its axis are the next steps they will take.

Materials:

- a globe
- models of the earth, moon, and sun
- pictures or posters of the daytime and nighttime sky
- drawing paper and crayons/markers

Earth Science Reference Software

- *Science Blaster Jr.* (Davidson)
- *School House Science* (Davidson)
- *Sammy's Science House* (Edmark)

Productivity Software

- *Draw* or *Paint*
- *The Writing Center*
- *Kid Pix Studio*

Procedure

Into: Before the Computer

- After looking at and discussing pictures depicting day and night skies, have the children draw pictures of what they see at night and what they see during the day.
- If appropriate, they can label their objects or write sentences describing them.
- Choose a productivity software program.

*For this example *Paint* was used.

Through: On the Computer

- Have the children divide their screens in two.
- Label one side "Day Sky" and the other "Night Sky."
- Have them use the drawing tools to draw what they see during the day and at nighttime.
- They can label their drawings or write sentences to explain.
- Print out their pictures and paste them onto large blue construction paper.
- Fold in half and label the front for a book about the cycle of day and night.

DAY AND NIGHT SKY (cont.)

Beyond: Extra Activities

- Create a three-card slide show with a title card, a "Day Sky" card, and a "Night Sky" card.
- The children can record what they see.
- Begin a cycles bulletin board with printouts of their pictures.

CYCLES

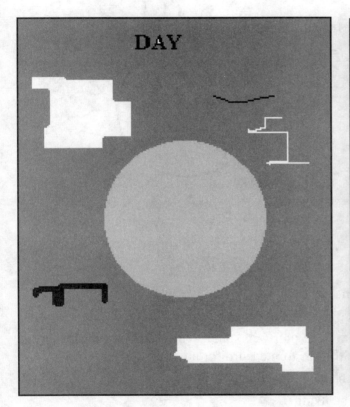

During the daytime I see the sun and clouds in the sky. Sometimes I see birds flying and lighting.

At night I see the moon and the stars.

Integrated Lesson Plans

SHADOWS

Measuring their shadows at various intervals during the day gets the children moving and outside. Working with a partner both on and off the computer will give them an opportunity at cooperative learning as they learn about the earth's movement around the sun.

Materials:

- models of the earth, moon, and sun
- tape measure or meterstick
- drawing paper and crayons/markers

Earth Science Reference Software
- *Science Blaster Jr.* (Davidson)
- *School House Science* (Davidson)
- *Sammy's Science House* (Edmark)

Productivity Software
- *The Graph Club*
- *Microsoft Works Spreadsheet*
- *Kid Pix Studio*

Procedure:

Into: Before the Computer

- Give each pair of students the worksheet on the next page.
- Take them outdoors on a sunny day to measure their shadows every hour, beginning at 9:00 A.M.
- Have each pair mark their spot where one from each pair will stand while their partner measures the shadow. (The same person in each duo will have to measure for accuracy.)
- They record their measurements on the worksheet.
- Decide on the productivity software you will be using.

Integrated Lesson Plans

SHADOWS (cont.)

Shadows Worksheet

Names: _____ Date: _____

Directions: Measure carefully and record your measurements here next to the time.

Time	Centimeters
9:00	_____
10:00	_____
11:00	_____
12:00	_____
1:00	_____
2:00	_____
3:00	_____

We measure our shadows as the earth turns.

Integrated Lesson Plans

SHADOWS (cont.)

Microsoft Works Spreadsheet was used for this activity.

Through: On the Computer

- The children will create a spreadsheet inputting 9, 10, 11, 12, 1, 2, 3 in the A column.
- They will put their measurements in centimeters in the B column.
- Highlight both columns to create a bar graph.
- Under Advanced Options, select Category Labels.
- Under Edit, Titles, the chart title is "Shadows"; subtitle, "by___"; horizontal axis, "9 A.M. until 3 P.M.", vertical axis, "Centimeters."
- Print out two copies of each chart, one for each partner.

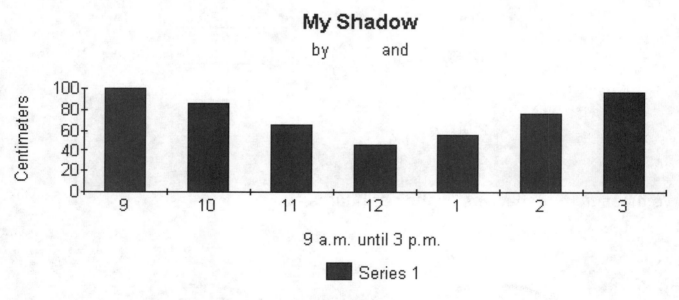

(*Microsoft Works*, Microsoft Corporation)

Beyond: Extra Activities

- If time permits, allow partners to switch roles and repeat the activity.
- Allow partners to create a slide show demonstrating the step-by-step procedure to measure their shadows.
- Use this activity as a springboard to discuss the earth's spinning on its axis as an explanation of day and night.
- Also address earth's tilt to explain the lengths of days and nights.

CONSTELLATIONS AND MYTHS

This is a fine topic on which to write across the curriculum. The constellations and their myths have excited stargazers over the centuries—whether as a guide for early travelers or as a destination for present-day astronomers.

Materials:

- pictures of constellations and constellation charts, if available
- drawing paper and crayons/markers

Productivity Software
- *The Amazing Writing Machine*
- *Kid Pix Studio*
- *Paint*
- *Microsoft Works*

Into: Before the Computer

- Have the children look at the night sky for a homework assignment. They could illustrate some of the stars they see.
- Take a field trip to a local university's planetarium.
- Discuss the night sky and the constellations.
- Help the children make their own constellation gazer with a recycled oatmeal container painted black and the bottom full of holes.
- Read a few constellation myths from your favorite sources. *Keepers of the Earth* contains myths explaining various mysteries in nature.
- Help the children use heavy white crayon to draw their constellation; then use a watercolor wash with just a splash of black to cover the paper.
- Give them examples and practice writing with a rebus.

Through: On the Computer

- Using graphics and clip art or the paint tool, have the children write a rebus myth to accompany their constellation watercolor.
- Print out the rebus and hang it on the bulletin board with their watercolors.

Beyond: Extra Activities

- Use the productivity software to create the constellation.
- Create a slide show with all of the children's myths. They can create the constellation on the slide and record their myths.

Integrated Lesson Plans

CONSTELLATIONS AND MYTHS (cont.)

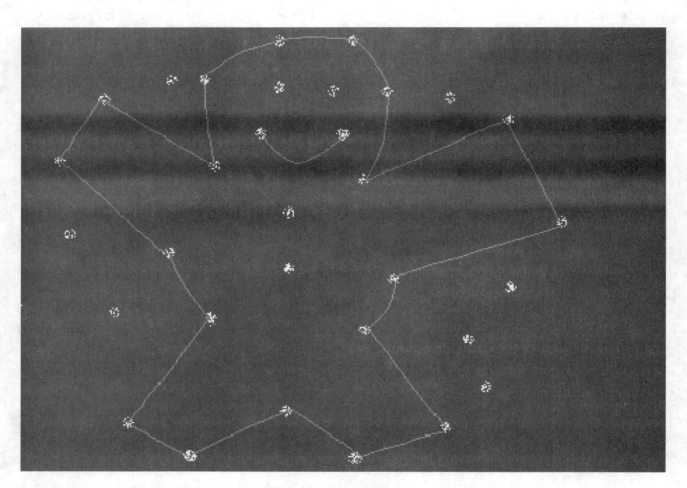

Why the Teddy Bear Smiles

One day while the teddy was riding his bicycle, he heard some music. This made his heart happy. He loved music. So he rode his bicycle home and all eyes were on him. And that is why the teddy is happy when you see him.

Integrated Lesson Plans

THE MOON'S CYCLE

When exploring the night sky, the children discover that the moon takes various shapes and sizes. How this occurs is not an easy concept for young children to totally master, but they will agree that they can observe the moon at four distinct cycles at fairly regular intervals.

Materials:

- pictures/posters of the moon and its phases
- moon calendar
- Science Activity Form, page 88

Procedure:

Productivity Software
- *Print Shop*
- *Draw* or *Paint*
- *Microsoft Publisher*
- *Kid Pix Studio*

Into: Before the Computer

- After studying and discussing the pictures of the moon's phases, the students will be asked to hypothesize the next month's moon calendar.
- They will use the Science Activity Form on the next page to predict the moon's phases for the next month.
- Then they will create a calendar of the actual phases of the moon for the next month to determine the accuracy of their predictions.
- Choose the productivity software for their moon calendar. *Print Shop* and *Microsoft Publisher* actually have calendar templates.

*Paint was used for this activity, drawing our own calendar and moon phases.

Through: On the Computer

- After making their predictions on the Science Activity Form, the children will draw a calendar on the computer.
- They have to know how many days are in the particular month. I do this activity in December since the children can take their printouts of their calendars wherever they go on holiday.
- They will draw the actual moon phases right on their printouts so that when they return to school after holiday, they can assess the assignment.

© Teacher Created Materials, Inc.

Integrated Lesson Plans

THE MOON'S CYCLE (cont.)

Science Activity Form

Name:_____ Date:_____

Title: _____

Question: _____

Hypothesis: _____

Procedure:

Materials:_____

_____

1. _____
2. _____
3. _____
4. _____

Results: _____

_____

_____

_____

Conclusion: _____

_____

_____

_____

_____

Integrated Lesson Plans

THE MOON'S CYCLE (cont.)

Science Activity Form

Name:_____ Date:_____

Title: _____The Moon's Phases_____

Question: ____When will we have a full moon next month?_____

Hypothesis: _____I predict that we will have a full moon on December 24.____

Procedure:

Materials:_____a moon calendar that I created on the computer_____

_____

1. _Make a moon calendar with your prediction for the moon's phases._
2. _Check the sky every night for the entire month, and draw the shape_
   onto your moon calendar.
3. _____
4. _____

Results: ___The full moon took place on December 23._____

_____

Conclusion: _I was only one day off._____

_____

© Teacher Created Materials, Inc.

Integrated Lesson Plans

THE MOON'S CYCLE (cont.)

Moon Calendar Prediction

Name: _____

DECEMBER, 1998

Sun.	Mon.	Tues.	Wed.	Thurs.	Fri.	Sat.
		1	2	3	4	5
6	7	8	9 ●	10	11	12
13	14	15	16  ◠	17	18	19
20	21	22	23	24  ○	25	26
27	28	29	30	31 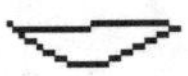 ◡		

#2427 Integrating Technology into the Science Curriculum   90   © Teacher Created Materials, Inc.

Integrated Lesson Plans

THE CYCLE OF THE SEASONS

Depending on geographic location, each of the seasons may be very different from one another. For some locations seasons do not differ much at all. Younger children begin to learn that the tilt of the earth's axis and its revolution around the sun are the causes for the cycle of the seasons.

Materials:

- models of the earth and sun
- books demonstrating how the tilt of the earth's axis causes the changing seasons
- pictures/posters of the changing seasons
- drawing paper and crayons/markers

Earth Science Reference Software
- *School House Science* (Davidson)
- *Sammy's Science House* (Edmark)

Productivity Software
- *Draw* or *Paint*
- *Kid Pix Studio*
- *HyperStudio*

Procedure:

Into: Before the Computer

- After looking at pictures of the four seasons, allow the children to manipulate models of the sun and earth.
- One child holding a flashlight is the sun; another holds the tilted earth and slowly revolves around the "sun" while slowly rotating the "earth."
- The earth stops at four locations while the children discuss the sun's rays on North America at each stop.
- Since North America receives the sun's rays for a longer period of time in 24 hours for three months of the year, it is warmer at that time of the year.
- During that same period Australia is tilted away from the sun; therefore, Australia receives less sunlight in 24 hours. It is colder at that time of the year.
- When it is summer in North America, it is winter in Australia.
- Divide the children into four groups, one group for each season.
- Allow each group time to discuss what it would look like in the temperate zone during their season. Then each child draws a picture of that season.

Integrated Lesson Plans

THE CYCLE OF THE SEASONS (cont.)

Through: On the Computer

- Depending on computer access, allow each child computer time to use a productivity software to create a picture of his/her season on the computer. A second picture shows the placement of the sun and earth for the season.
- Print out the pictures. Then group the children in fours, one from each season, to describe his/her season to his/her group. See samples of computer printouts below and on the following pages.
- Use the self-assessment on page 95 to determine the comfort level for each child at the computer.

Beyond: Extra Activities

- Combine printouts of the four seasons onto a large poster entitled "The Cycle of the Seasons."
- Allow each child to complete a multimedia card *(HyperStudio,* for example) to create a stack of the same title.
- Take digital pictures of each child for the title card so that when the child's picture is clicked on, his/her season comes next..

Spring

Autumn

Integrated Lesson Plans

THE CYCLE OF THE SEASONS (cont.)

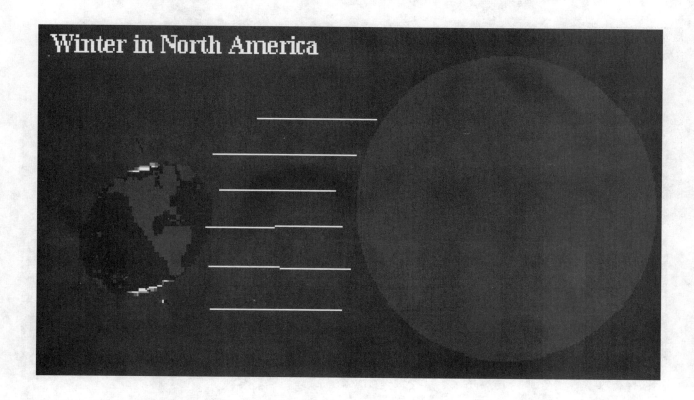

© Teacher Created Materials, Inc. 93 #2427 Integrating Technology into the Science Curriculum

Integrated Lesson Plans

THE CYCLE OF THE SEASONS (cont.)

THE CYCLE OF THE SEASONS (cont.)

PROJECT ASSESSMENT FORM

Name _____

Date _____

This is how I felt about the project: _____

_____

_____

_____

                                                        Yes   No

	Yes	No
I followed directions.	_____	_____
I used the computer well.	_____	_____
I planned ahead.	_____	_____
I can help others with this project.	_____	_____

Integrated Lesson Plans

THE WATER CYCLE

Depending on the age level, a lesson on the water cycle can be as simple or as complex as desired. I have observed a second-grade class sing a song about the water cycle with wonderful choreography created by the students. Creating computer-generated, three-dimensional pictures can be appropriate for all primary children.

Materials:

- models of the water cycle (commercially or student made)
- pictures/posters of the water cycle
- glass jar with lid, warm water, ice cubes
- spray bottle of water
- smooth-surface ramp
- drawing paper and crayons/markers

Water Cycle Reference Software
Water Planet (Science for Kids)

Productivity Software
- *Draw* or *Paint*
- *Kid Pix Studio*

Procedure:

Into: Before the Computer

- As a whole-class activity with the children performing, warm water is poured into a glass jar.
- Place ice cubes on top of the inverted lid which is placed on the jar opening.
- When "clouds" have formed, after about five minutes, remove the cubes and lid to discover condensation on the bottom of the lid.
- After ten minutes the "rain" will fall.
- To demonstrate run-off, have a student use a spray bottle of water and squirt a smooth ramp.
- Another student should run his/her finger across the droplets to demonstrate how the heavier droplets will run down the hill.
- Discuss with the children that what they have witnessed is similar to how the water cycle works.
- Give the children the planning sheets on the following page to draw their own water cycles. They may work with a partner to complete their water cycles.
- Depending on grade level, they will simply draw the picture, add words, or write accompanying sentences.
- Choose a productivity software to create the student pictures on the computer.

THE WATER CYCLE (cont.)

The Water Cycle Planning Sheet

Directions: Draw a picture of your water cycle here:

Directions: Use the words below to tell how the water cycle works.

      evaporation   condensation   precipitation

The sun causes _____. Puddles get smaller and disappear when this happens.

_____ is when the evaporated water forms droplets which join together to become clouds.

When the clouds get full of droplets, they cannot hold any more water. Then _____ occurs, and rain, hail, freezing rain, or snow falls from the clouds.

Integrated Lesson Plans

THE WATER CYCLE (cont.)

*For this example *Kid Pix Studio* was used.

Through: On the Computer

- It is ideal to have the children work with partners when creating on the computer. They are able to assist each other both with the software and with the final product.
- They will use their planning sheets to create a computer-generated water cycle. (See the sample on page 99.)
- They can use the *Kid Pix* stamps or the tools to draw the parts of the water cycle.
- Depending on their grade level, they can label their pictures or describe how the water cycle works below their pictures.
- Allow time for the partners to present their water cycles to the rest of the class.

Beyond: Extra Activities

- Print the pictures, cut them out, and paste them onto heavy construction paper to create three-dimensional water cycles. (See the "Safety" lesson on pages 26–28 for three-dimensional instructions.)
- Have groups of three children create a water cycle slide show. Each of the children's slides will demonstrate evaporation, condensation, or precipitation. Each group of three presents to the rest of the class.
- Another alternative would be to put six children in each group. While three create the slide show, the other three create a song and movement to accompany the slide show.

THE WATER CYCLE (cont.)

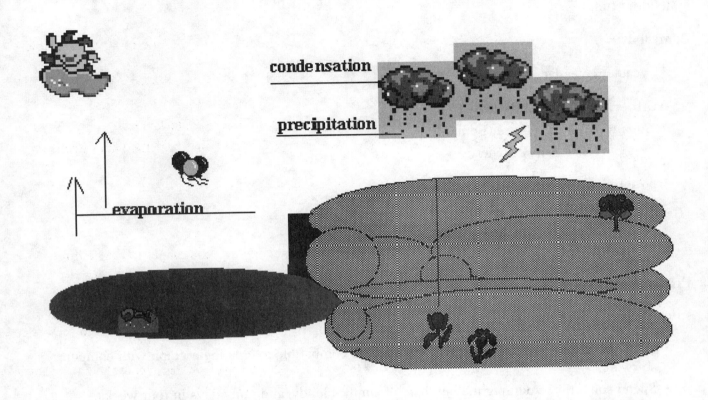

The Water Cycle

When it is warm, the sun helps to evaporate the water in puddles, rivers, and oceans.

The evaporated water forms droplets around dirt particles.

When many droplets come together, they form a cloud.

When the cloud is full (saturated), precipitation occurs in the form of rain, hail, freezing rain, or snow.

The falling rain runs into puddles, streams, rivers, and oceans.

The water cycle will begin all over again.

Integrated Lesson Plans

WHAT'S THE WEATHER?

Children learn to keep track of the weather on their weather calendars as early as preschool. By teaching them to graph this information also at an early age, they find that reading and interpreting graphs are fun.

Materials:

> weather calendar

Weather Reference Software
- *Sammy's Science House* (Edmark)
- *School House Science* (Sierra)

Productivity Software
- *Draw* or *Paint*
- *Kid Pix Studio*
- *The Graph Club*

Procedure:

Into: Before the Computer

- For your weather calendar use symbols that are compatible to creating a graph with computer software.
- Spend some time counting the numbers of sunny, cloudy, and rainy days in four weeks.
- Discuss the weather cycle and water cycle.
- Choose a graphing software.

The Graph Club was used for this activity.

Through: On the Computer

- Choose any four weeks during which the class will keep track of the weather.
- Graph the number of sunny, rainy, cloudy, etc., days. (See the samples on pages 101 and 102.) The season and the school's geographic location will determine the pictures on the graph.
- Divide the class so that each student will work with a group for graphing a particular week.
- Once started this can be an ongoing project with the preceding week's graph displayed.

Beyond: Extra Activities

- Use draw software to show how the children dress for the various kinds of weather.
- Create a graph depicting the "Favorite Season" choices of the students.

Integrated Lesson Plans

WHAT'S THE WEATHER? *(cont.)*

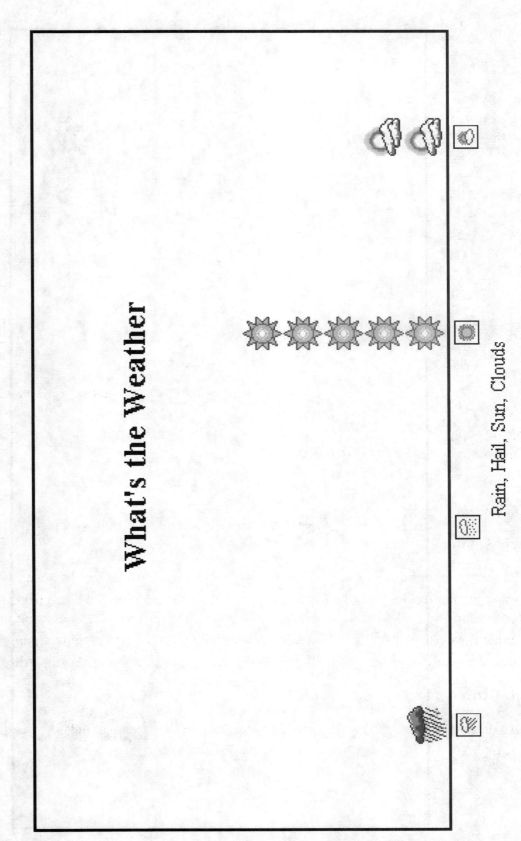

Integrated Lesson Plans

WHAT'S THE WEATHER? *(cont.)*

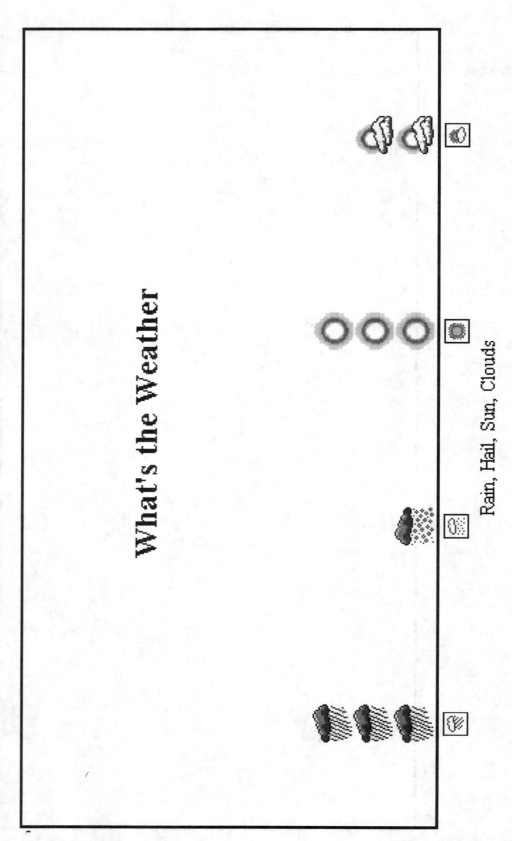

Integrated Lesson Plans

AIR TEMPERATURE

The GLOBE (Global Learning and Observations to Benefit the Environment) Program was initiated by Vice President Al Gore and is open to school children all over the world. The program invites students to collect valuable environmental data and perform experiments under the direction of educators and scientists. The information is sent to a central database and used by researchers to study trends in global change and environmental degradation. Keeping a record of the air temperature on a daily basis is part of this program. After monitoring air temperature and precipitation, along with observation of cloud types, you might be interested in becoming a GLOBE teacher for your school. The URL is listed on pages 16 and 104. You and your students might enjoy taking a look at this address to find weather patterns all over the world. Students in other countries can even be e-mailed via this URL.

Materials:

- an accurate thermometer (possibly one that depicts minimum, maximum, and current temperatures)
- Air Temperature Recording sheet (page 105)
- texts and other reference material on weather stations and gauges

Weather Reference Software
- *Sammy's Science House* (Edmark)
- *School House Science* (Sierra)

Productivity Software
- *Draw* or *Paint*
- *Kid Pix Studio*
- *The Graph Club*
- *Microsoft Works Spreadsheet*

Procedure:

Into: Before the Computer

- Bring in a pilot or other worker who depends on weather predictions to speak with the children about the importance of understanding weather patterns.
- Give the children a great deal of practice reading a thermometer.
- Decide whether to use Celsius or Fahrenheit and explain the difference.
- Have the children take turns accurately recording the air temperature for several weeks.
- Check the local newspapers to determine how close they are to readings at local weather stations.
- Determine which graphing software to use.

Integrated Lesson Plans

AIR TEMPERATURE (cont.)

Microsoft Works Spreadsheet was used for this activity.

Through: On the Computer

- Create a spreadsheet by using the readings on the weather recording chart on page 105. (See the sample spreadsheet on page 106.)
- Create a line graph depicting a week's air temperature. (See the sample graphs on page 107.) Explain that a line graph is best for this activity since it measures change over a period of time.
- Print out new graphs weekly to see the weather pattern for your area.
- Compare this to weather patterns all over the world.

Beyond: Extra Activities

- Make a "What's the Weather?" bulletin board with computer-generated charts and pictures showing weather all over the world.
- Have the children create weather-related poems on the word processor. Use a paint program or clip art to create pictures that will accompany the poems. Make a "Weather Poems" book which includes each student's poem.
- Find a school across the world to e-mail and send pictures and stories about your school.

Internet Connection:

http://globe.fsl.noaa.gov/ **G**lobal **L**earning and **O**bservations for a **B**etter **E**nvironment: GLOBE students all over the world are taking environmental measurements and sharing this information with scientists and technologists via the Internet.

Integrated Lesson Plans

AIR TEMPERATURE (cont.)

Air Temperature Recording Sheet

Air Temperature Chart—Max/Fahrenheit					
Day	Date	Temp.	Day	Date	Temp.
Average Monthly Temperature					

AIR TEMPERATURE (cont.)

Sample Air Temperature Spreadsheet

Air Temperature Chart–Max/Fahrenheit					
Day	Date	Temp.	Day	Date	Temp.
Wed.	1	72	Wed.	15	59
Thurs.	2	73	Thurs.	16	57
Fri.	3	69	Fri.	17	56
Sat.	4	68	Sat.	18	58
Sun.	5	68	Sun.	19	60
Mon.	6	71	Mon.	20	62
Tues.	7	74	Tues.	21	62
Wed.	8	78	Wed.	22	65
Thurs.	9	81	Thurs.	23	68
Fri.	10	80	Fri.	24	72
Sat.	11	82	Sat.	25	75
Sun.	12	76	Sun.	26	79
Mon.	13	75	Mon.	27	77
Tues.	14	75	Tues.	28	79
Average Monthly Temperature					70

AIR TEMPERATURE (cont.)

Sample Air Temperature Graphs

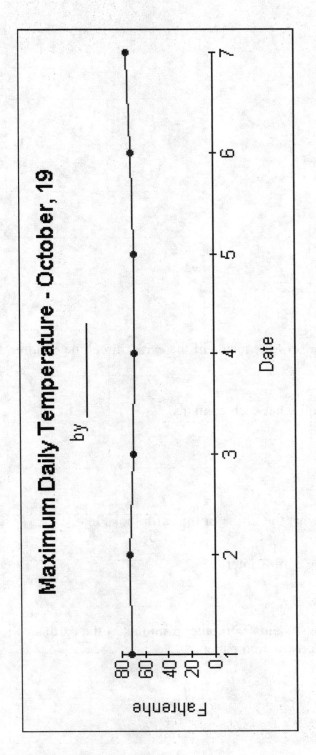

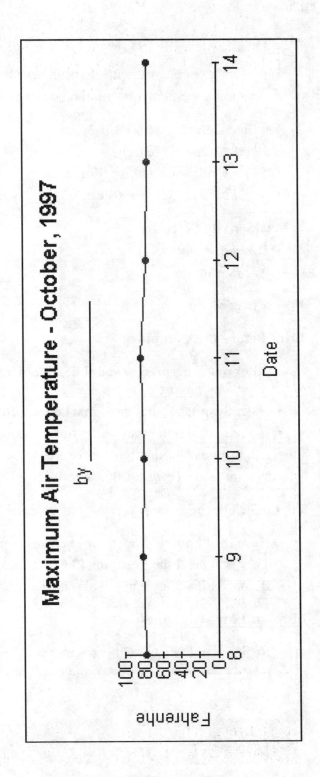

Integrated Lesson Plans

INSIDE EARTH

Fascination with Earth—its beginning, its composition, its changes—motivates children of all ages. Begin with drawing the earth's crust, mantle, and core; continue with studying fossils and dinosaurs, and children are captivated. Their drawings can be realistic, abstract, or surrealistic—this is their choice or the teacher's. Have fun!

Materials:

- posters/pictures of the inside of the earth
- drawing paper, crayons, markers

Weather Reference Software
- *Earth Science* (Clearvue)
- *Sammy's Science House* (Edmark)
- *School House Science* (Sierra)

Productivity Software
- *Draw* or *Paint*
- *Kid Pix Studio*

Procedure:

Into: Before the Computer

- After much discussion and time to look at pictures of the inside of the earth, direct the children to draw a picture of it.
- Have them label the crust, mantle, and core.
- If age appropriate, they can write sentences to tell what each contains.
- Choose your productivity software.
- We are using *Paint* for this activity.

Through: On the Computer

- Have the children draw their interpretations of what the inside of the earth looks like.
- They can label the crust, mantle, and core.
- If appropriate they can tell about each part of the earth's interior.

Beyond: Extra Activities

- Begin a book about "Changes on Earth" and use students' computer printouts of the earth's interior for the cover. Paste the printouts onto construction paper.

INSIDE EARTH (cont.)

Inside the Earth

by _____

We live on the earth's crust. It is the outside of the earth, and it is about six kilometers thick.

No one has drilled through all of the earth's crust to get to the mantle. The mantle is the thickest part of the earth. Most scientists believe the mantle is made of very hot rock.

The core is probably liquid since we believe that it is very, very hot iron.

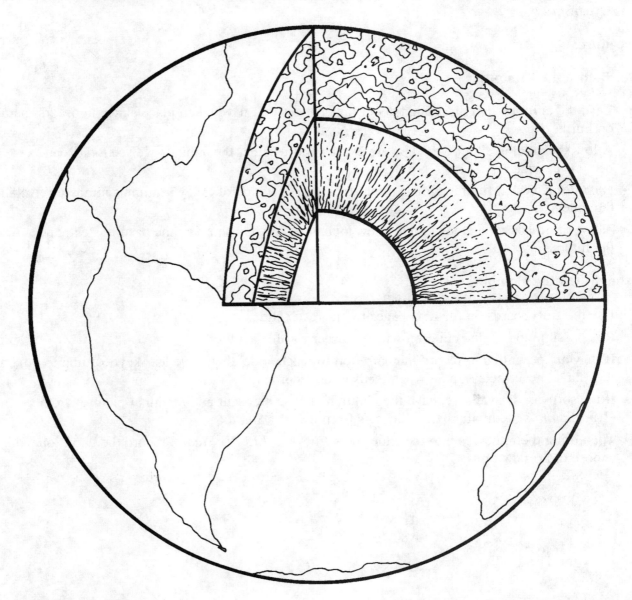

Integrated Lesson Plans

PAST, PRESENT, FUTURE

Young scientists living in the present can imagine what happened in the past and what the future will look like. What makes this assignment more interesting is creating a slide show to present to others.

Materials:

big picture books/posters about dinosaurs

Dinosaur Reference Software
- *Learn About Dinosaurs* (Sunburst)

Productivity Software
- *Draw* or *Paint*
- *Kid Pix Studio*
- *HyperStudio*

Procedure:

Into: Before the Computer

- Lead the children in a discussion about the changes that have taken place since dinosaurs roamed on Earth.
- Allow speculation about why they are extinct and whether the animals we see today can become extinct.
- Next discuss what future animals might look like or which of today's animals might survive all of the others.
- All of this discussion leads into the development of a slide show on the past, present, and the future.

Through: On the Computer

- Use the storyboards on the next pages to plan each slide.
- It is possible to pair the children with a partner for this activity.
- If they are permitted to record the narration for each slide, they have less keyboarding to do, but they must, nevertheless, plan what needs to be recorded.
- If the software is available, have the children choose dinosaur and mammal pictures from *HyperStudio's* media library and insects from *Kid Pix Studio*.
- Then bring their choices into the slide show trucks in *Kid Pix Studio*, record the narration, and choose their transitions.

Integrated Lesson Plans

PAST, PRESENT, FUTURE (cont.)

Storyboard Planning

Past	**Past**

Put your drawings of dinosaurs in these boxes labeled "PAST" and then tell about the dinosaurs of the past and how you think they became extinct.

_____
_____
_____
_____
_____
_____
_____

© *Teacher Created Materials, Inc.*   111   #2427 *Integrating Technology into the Science Curriculum*

Integrated Lesson Plans

PAST, PRESENT, FUTURE (cont.)

Storyboard Planning

Present	**Present**

Draw pictures of animals of today. Tell which of these animals might become extinct. Why might that happen?

_____
_____
_____
_____
_____
_____
_____

Integrated Lesson Plans

PAST, PRESENT, FUTURE *(cont.)*

Storyboard Planning

Future

Future

Draw pictures of the animals that might live in the future. Tell why you believe that they will survive.

_____
_____
_____
_____
_____
_____
_____

© Teacher Created Materials, Inc. 113 #2427 Integrating Technology into the Science Curriculum

Integrated Lesson Plans

PAST, PRESENT, FUTURE (cont.)

Slide Show Sample

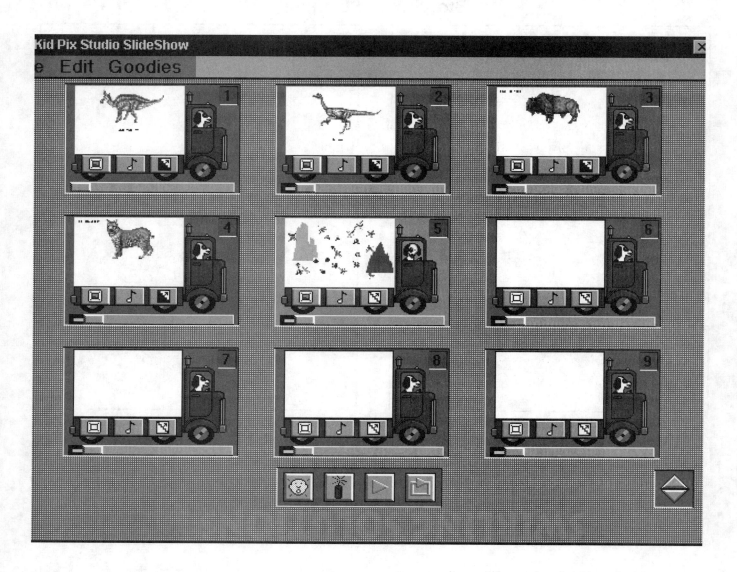

ROCKS WORD SEARCH

Rocks, minerals, crystals—there is lots of investigation to be done on this topic. As a culminating activity, the children make their own word searches on the computer.

Materials:

- texts, pictures, posters depicting minerals, crystals, and three types of rocks: igneous, sedimentary, and metamorphic
- results of activities done by you and children, demonstrating hardnesses of various minerals
- an easy crystal-forming activity: Dissolve ordinary table salt in a glass jar of warm water and keep adding salt until it is totally saturated. Allow the opened jar to just sit and observe crystals as the water evaporates.

Weather Reference Software
- *Science II* (Advanced Learning Systems)
- *Rocks & Volcanoes* (Queue, Inc.)

Productivity Software
- *Word Search Puzzles* (Gamco)
- *Wordsearch Creator* (Centron)
- *Worksheet Magic* (Teacher Support)
- any spreadsheet software

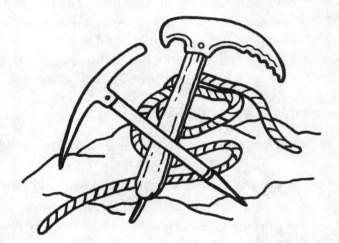

Procedure:

Into: Before the Computer

- Explore the world of minerals, crystals, and rocks with the children in order to increase their vocabularies in this area.
- Allow each child to use a dime-store magnifying glass to closely observe the various types of rocks.
- Categorize the rocks according to their characteristics.
- Do a whole-class brainstorming session to get as many vocabulary words on the topic as possible. Put all suggestions on the board.

Integrated Lesson Plans

ROCKS WORD SEARCH (cont.)

Through: On the Computer

- There are many word search software products, but the one used here is a spreadsheet format.
- Have the children work with a partner and use the vocabulary that the whole class has brainstormed to complete their word searches.
- If it is necessary to do the computer work on another day, simply do the brainstorming on poster paper or bulletin board paper so that it can be visible to the computer work station.
- The children will take turns putting the words into the word search; one types as the other spells the word. Have them include a word list of all words that are in their searches.
- In order to put the words diagonally, show them how to type a letter on one line and then move over one space on the next line. The first example on the next page shows only the vocabulary words; the following page shows a completed word search. It may be reproduced to give the children practice in using word searches.
- The children can print out multiple copies of their word searches, or they can be copied so that they can exchange them with their classmates.

Beyond: Extra Activities

- Divide the class into the three rock types and have each group include only those rocks that are classified as their type in their word searches.
- Instead of a word search, create rock and mineral crossword puzzles.
- Create a rock/mineral game board in which correct answers to questions written on index cards permit the moves. Use stones for moving pieces. The game board could be a search for the gold mine which is the final move.

ROCKS WORD SEARCH (cont.)

Sample Word Search

Rocks and Minerals Word List

minerals	salt	quartz	graphite	rock
diamond	talc	calcite	igneous	shale
granite	magma	sedimentary	coal	
limestone	sandstone	metamorphic	chalk	
slate	marble	fossil	crystals	pumice

Rocks and Minerals Word Search

m	e	t	a	m	o	r	p	h	i	c	r	q	l	s	o
a	d	q	u	s	s	e	d	i	m	e	n	t	a	r	y
p	i	q	u	h	c	r	y	s	t	a	l	s	p	o	s
u	a	l	i	a	a	r	e	n	i	m	i	o	n	e	l
m	m	i	t	l	r	u	o	s	t	a	s	a	l	t	a
i	o	m	a	e	r	t	a	l	c	g	s	g	m	a	r
c	n	e	m	a	g	t	z	a	h	m	o	u	i	r	e
e	d	s	a	r	o	c	k	t	a	a	f	e	l	t	n
d	e	t	r	m	d	b	a	e	l	t	n	o	p	v	i
w	a	o	b	o	a	c	t	p	k	o	n	t	q	u	m
k	b	n	l	l	r	a	o	e	t	i	n	a	r	g	e
l	b	e	e	a	l	u	m	s	t	e	e	l	t	z	y
a	z	d	l	o	n	i	d	c	a	l	c	i	t	e	m
h	e	l	i	c	u	n	m	g	r	a	p	h	i	t	e
c	l	g	a	s	a	m	e	t	a	l	s	g	o	o	d
g	o	l	d	s	o	i	l	s	u	o	e	n	g	i	m

Integrated Lesson Plans

ROCKS WORD SEARCH (cont.)
Sample Word Search

Rocks and Minerals Word Search Answer Key

m	e	t	a	m	o	r	p	h	i	c			
	d	q		s	s	e	d	i	m	e	n	t	a	r	y
p	i		u	h	c	r	y	s	t	a	l	s		s
u	a	l		a					m	i	o	n	e	l
m	m	i		l	r		s		a	s	a	l	t	a
i	o	m		e	r	t	a	l	c	g	s	g	m	a	r
c	n	e	m		g	t	z	a	h	m	o	u	i	r	e
e	d	s	a	r	o	c	k	t	a	a	f	e		n
		t	r		d	b	a	e	l		n			i
		o	b				k	o						m
k		n	l	l	r	a	o	e	t	i	n	a	r	g
l		e	e	a	l	u	m	s					
a			o	n	i	d	c	a	l	c	i	t	e
h			c		n		g	r	a	p	h	i	t	e
c				a									
			s			s	u	o	e	n	g	i	

Integrated Lesson Plans

DINOSAUR DATABASE

The study of dinosaurs is always motivating for primary age children. Classifying dinosaurs is an interesting way to introduce students to creating databases. Once they have listed the various characteristics denoting similarities and differences of these extinct animals, they are ready to begin working on their databases.

Materials:

- text, books, posters, charts with pictures of dinosaurs, their choices for food, and their sizes
- Dinosaur Planning Sheet on page 121

Dinosaur Reference Software

- *3-D Dinosaur Adventure* (Knowledge Adventure)
- *Dinosaur Days* (Toucan)
- *Dinosaur Kids* (Nordic)

Productivity Software

- database software

Procedure:

Into: Before the Computer

- The children can be grouped in various ways for this lesson. They can be in groups according to dinosaur categories: meat eaters, plant eaters, etc. Or each group of three or four students can do an independent research on any six dinosaurs.
- Each group uses the dinosaur planning sheet provided with this lesson to list the dinosaurs and tell characteristics about each. Add any fields deemed appropriate for the age level.
- Each group can have resident artists who will draw pictures to accompany the database.
- One person in the group can be the recorder who fills out the planning sheet or the children can take turns with this part of the job.
- The one in charge of waste management makes sure all supplies are put in their places and all paper not used is recycled.

Through: On the Computer

- If this is the first time for creating a database, be sure to use the tutorial. The screen shot on page 122 is just a sample.
- The children need to take their planning sheets to the computer.
- It is wise to allow them to take turns inputting data into the computer. Perhaps each student can complete all of the information for one dinosaur.
- The first field should name the dinosaur, but the other fields can be whatever you deem appropriate for the students' age level. Add fields to the dinosaur planning sheet on the next page.

© Teacher Created Materials, Inc.

Integrated Lesson Plans

DINOSAUR DATABASE (cont.)

Through: On the Computer *(cont.)*

- There is no need to be concerned with alphabetizing when inputting the dinosaur names, since sorting can be done automatically with the sorting tool. In fact, it is fun for the children to see how they can sort by any field: name, size, etc. The computer understands the less than and greater than symbols, too.
- Have each group print out their databases and share their results with the rest of the class.

Beyond: Extra Activities

- Create a "Dinosaur Characteristics" bulletin board. With each database include printouts of drawings the children have created on the computer.
- Create a three-dimensional dinosaur shadowbox. (See *Safety* on page 26 for three-dimensional instructions.)
- Create a dinosaur stack using *HyperStudio*. The children can each create three cards depicting the characteristics of their dinosaur.
- Have the children create dinosaur parts similar to those on page 123 with a draw/paint software. Then others in the class can cut up the parts and put together an imaginary dinosaur.

Integrated Lesson Plans

DINOSAUR DATABASE (cont.)

Dinosaur Planning Sheet

Name	Length	Food

Integrated Lesson Plans

DINOSAUR DATABASE (cont.)

Sample Dinosaur Database

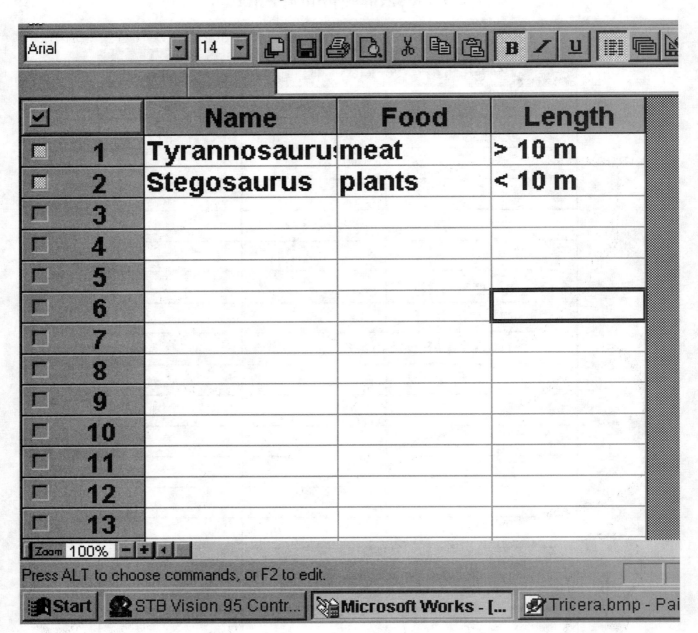

(*Microsoft Works*, Microsoft Corporation)

DINOSAUR DATABASE (cont.)

Create an Imaginary Dinosaur

Directions: Cut out the parts and paste some together onto construction paper to create an imaginary dinosaur.

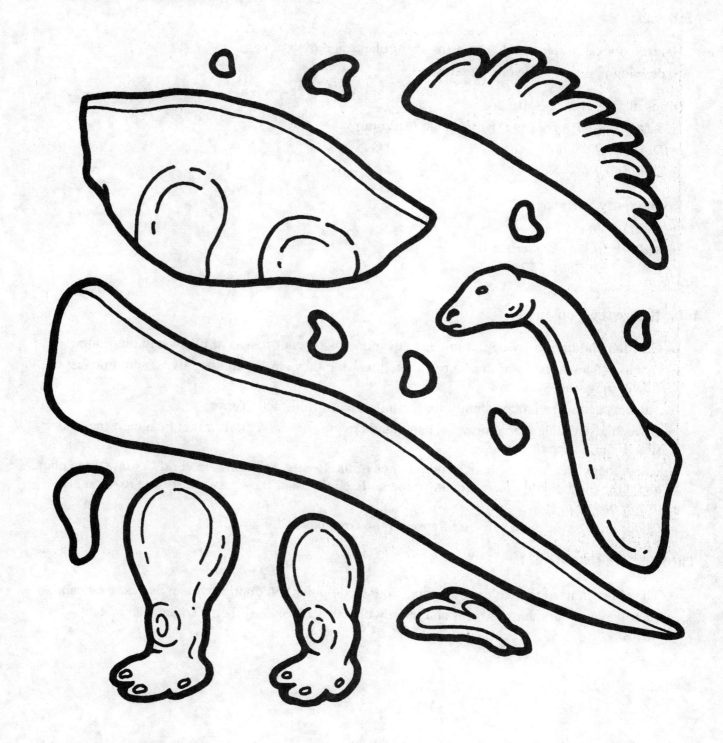

Integrated Lesson Plans

MATCHING FOSSILS

Scientists use the evidence garnered from fossils to determine many facts about the dinosaurs. The dinosaur's teeth tell what it ate; the size and shape of its legs determine its speed. Now the children will have the opportunity to draw fossils of plants and dinosaurs and see if their classmates can match the two.

Materials:

- texts, books, poster, charts about fossils of plants and dinosaurs
- drawing paper and crayons/markers

Dinosaur Reference Software

- *3-D Dinosaur Adventure* (Knowledge Adventure)
- *Dinosaur Days* (Toucan)
- *Dinosaur Kids* (Nordic)

Productivity Software

- *Paint* or *Draw*
- *Kid Pix Studio* (Broderbund)

Procedure:

Into: Before the Computer

- Give the children many opportunities to examine pictures of fossils of plants and dinosaurs.
- Discuss characteristics of dinosaurs that tell us what they ate, where they lived, and how fast or slowly they walked or ran.
- Students can work independently for the initial part of this activity.
- Give children each a large drawing paper and crayons or markers. It might be nice if they can stretch out on the floor.
- Allow them access to the books and pictures of the fossils, and ask each child to draw three plants and three dinosaurs on their drawing papers. Each dinosaur they draw must have its food also drawn on the page.
- Choose a draw or paint software to use on the computer.

Through: On the Computer

- The children need to take their drawings to the computer and duplicate them as best they can.
- Next they cut out their drawings and trade with partners to see if their partners can match their fossils.

MATCHING FOSSILS (cont.)

Beyond: Extra Activities

- Have the children draw a habitat on a bulletin board on which they paste their fossils.
- Put the names of dinosaurs on a large poster and have the children paste their drawings under the appropriate names.
- Make a bar graph showing the actual sizes of the various dinosaurs.

What Did It Eat?

Integrated Lesson Plans

SWIRLING SOLUTIONS

Differentiating between solutions and mixtures is made easy for primary students when you use everyday food products. Cut up various types of fruit to make a mixture of fruit salad; stir some syrup into milk to make a chocolate milk *solution*. The following activity evokes "oohs" and "aahs" because of the pretty results made with milk, food coloring, and dish soap. Besides the study of solutions, this lesson also lends itself to studying weather currents as the children observe the swirling liquid.

Materials:

- glass dishes or petri dishes (enough for half the class would be nice)
- small cubes to raise the dish
- flat mirrors (optional)
- quart of milk or half and half
- food coloring
- liquid dish soap
- drawing paper and crayons/markers

Productivity Software
- *Paint* or *Draw*
- *Kid Pix Studio* (Brøderbund)

Into: Before the Computer

- Allowing the children to work with a partner for this activity is beneficial since they will be able to get close up for observation. If fewer dishes are available, then group the children in threes.
- Also if flat mirrors are available to view the underside of the dish, that's also helpful. If not, the view from above is spectacular enough.
- Each duo will place four cubes under the sides of their dish and lay the mirror facing up under the dish.
- Allow a few partners to use clay to create continents or volcanoes or mountains on their dishes before adding the liquid. (This can also be done by the whole class the second time or not at all.)
- Pour the milk into the dish so that the surface is covered by about two mm. More does not matter; it just costs more.
- Drop a different color of food coloring in each of the four corners.
- This is when the swirling action begins. Place several drops of dish detergent in various places in the dish.
- After sitting back and observing for a few minutes, have the children sketch what they see.
- The swirling is caused by the interaction of the various molecules: soap, water, milk, and food coloring.

SWIRLING SOLUTIONS (cont.)

Into: Before the Computer (cont.)

- After drawing what they see, allow the children to use a spoon to stir the liquid so that they will see that in a *solution* the colors lose their individual characteristics and become a different-colored liquid unlike each individual color.
- Choose a draw or paint software to illustrate what is occurring.

Kid Pix Studio was used for this activity.

Through: On the Computer

- The children will use the circle tool throughout this activity.
- First, they make a large circle with black. This circle is not filled in so that it shows the white "milk."
- Next, staying with the circle tool, they choose one of the four colors. This time they choose the filled-in circle and make a large circle or oval in one corner of the "dish."
- With that same color, making only the outline not filled in, they make a large circle or oval inside the first one.
- Then going back to the filled-in circle, make another one on top of the second, and so on.
- Do the same thing in each corner with a different color until all four colors are used.
- The partners take turns so that each works with two colors.
- Print the results, put each on a different-color sheet of construction paper, and hang it on the bulletin board.

Beyond: Extra Activities

- To show the difference between *solutions* and *mixtures*, put two partner groups together and have each of the four children bring in a different piece of fruit.
- They can have the fruit cut up at home.
- Give each group of four a small bowl in which they place their fruit.
- This demonstrates a *mixture* in which the fruit is mixed together but each keeps its own properties.
- After drawing the result on the computer and printing the drawing, they can enjoy their fruit salad.

Integrated Lesson Plans

SWIRLING SOLUTIONS (cont.)

My Swirling Solution

#2427 Integrating Technology into the Science Curriculum   128   © *Teacher Created Materials, Inc.*

Integrated Lesson Plans

HEAT, LIGHT, AND SOUND

Energy is all around us, and we use it every day. This activity can be the initial lesson on energy in order to, first, motivate the children and, second, help the teacher assess what they already know about heat, light, and sound.

Materials:

- paper, pencil
- planning sheets, reproduced from following pages

Dinosaur Reference Software
- *Sammy's Science House* (Edmark)
- *Science II* (Advanced Learning Systems)

Productivity Software
- *Paint* or *Draw* software
- *Kid Pix Studio*

Procedure:

Into: Before the Computer

- With the children in groups of four or five, have them brainstorm as many objects that give off heat, light, and sound as possible. If their age is appropriate, one child can be recorder.
- Then the whole class shares their ideas as the objects are written on the board.
- Reproduce the planning sheets on the following pages so that the children can draw pictures of various objects for this activity.
- After drawing their pictures, have them meet again in their groups to tell why they chose their pictures to go with each form of energy.
- Choose a software for this activity.

Kid Pix Studio was used for this activity.

Through: On the Computer

- Using their planning sheets the children create three pictures on the computer—one for each form of energy. (See the samples on pages 133 and 134.)
- Each child chooses one form of energy to present to the class and tells why the pictures go with that form of energy.

Integrated Lesson Plans

HEAT, LIGHT, AND SOUND (cont.)

Planning Sheet

Heat

Integrated Lesson Plans

HEAT, LIGHT, AND SOUND (cont.)

Planning Sheet

Light

Integrated Lesson Plans

HEAT, LIGHT, AND SOUND (cont.)

Planning Sheet

<u>Sound</u>

HEAT, LIGHT, AND SOUND (cont.)
Samples

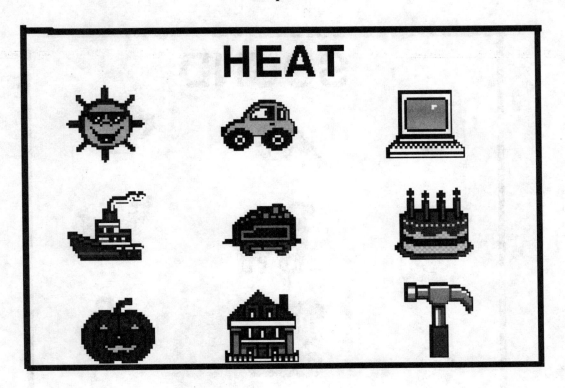

Integrated Lesson Plans

HEAT, LIGHT, AND SOUND (cont.)
Sample

Beyond: Extra Activities

- Create a slide show using the previous drawings. Each group can choose a few of each child's selections to include in the slide show.
- Actually record the sounds for the sound slide.
- Begin an "Energy" bulletin board with pictures for this activity as the first ones to include.
- If their age is appropriate, include the students' words or sentences to explain the drawings.

Integrated Lesson Plans

MACHINES HELP

There are many, many different kinds of machines. People change machines to make work easier and faster. A pencil can be sharpened with a small, hand-held pencil sharpener, with one fastened to the wall with a handle that is turned, or with an electric sharpener in which the pencil is simply held. People get from place to place in an automobile but faster in a train and fastest in an airplane. Perhaps one of your students may someday invent a machine that will help us.

Materials:

- pictures and posters of various types of machines, including simple machines and complex machines composed of many simple machines
- models of various types of machines, perhaps some that the children can bring from home
- *Tinker Toys®*, *Legos®*, or *Erector Sets®*
- pencil shapeners described above

Productivity Software
- spreadsheet software
- *Paint* or *Draw*
- *Kid Pix Studio*
- *The Graph Club*

Into: Before the Computer

- After studying pictures of various types of machines, allow the children time to create their own with the above-mentioned manipulatives. Their machines need to help save time.
- Have them tell what their machines can do and how they save time.
- Show the children the three types of pencil sharpeners and ask how they think each one saves time. (Before sharpeners a knife was used to whittle a pencil point.)
- Use the science activity form on page 88, if desired, so that they can hypothesize the number of pencils each sharpener can sharpen in one minute.
- Choose three children to sharpen the pencils as the class times them.
- Record their times.
- Choose a graphing software.

Integrated Lesson Plans

MACHINES HELP (cont.)

*The Graph Club was used for this activity.

Through: On the Computer

- Have the children work with a partner to show the results of the pencil sharpening activity by creating a graph on the computer.
- If you do not have *The Graph Club,* use a spreadsheet software which will create the graph for you.

Pencil Sharpeners	
What Kind?	Pencils sharpened in one minute
Hand Held	2
Rotary	5
Electric	10

(*The Graph Club,* Tom Snyder Productions)

EDUCATIONAL SOFTWARE DISTRIBUTORS

The following companies will respond to requests for catalogs containing educational software. When deciding on software for the classroom, be sure to express a desire to preview the product. Most companies will gladly allow a 30-day trial so that you can determine whether the software meets your needs.

Brøderbund
(800) 474-7740

Educational Software Institute
(800) 955-5570

Computer Plus
(800) 446-3713

Scantron
(800) 777-3642

Educational Resources
(800) 624-2926

SoftWareHouse
(800) 541-6078

Sunburst Communications
(800) 321-7511

PRODUCTIVITY SOFTWARE

Title: *The Amazing Writing Machine*
Publisher: Brøderbund

This easy-to-use and enjoyable program inspires children to write and illustrate their own books, journals, essays, letters, and poems.

Title: *Crossword Companion*
Publisher: Visions

Create a variety of looks with full word processing and a variety of print options, including different graphic and puzzle sizes, and pre-printing of selected letters or entire word lists.

Title: *The Graph Club*
Publisher: Tom Snyder Productions

This elementary graphing tool helps students learn to gather, sort, and classify information.

Title: *HyperStudio*
Publisher: Roger Wagner Publishing

HyperStudio is a hypermedia environment where non-technical users can bring graphics, sound, and text together in an interactive and interconnected way.

Title: *Kid Pix Studio*
Publisher: Brøderbund

This is a widely expanded CD-ROM version of the original *Kid Pix* plus a whole collection of multimedia features.

Title: *Microsoft Works*
Publisher: Microsoft Corp.

Here it is the powerful, versatile, and simple-to-use integrated software package for word processing, spreadsheets, database, and communications.

Title: *Word Search Puzzles*
Publisher: Gamco

Enter word lists and then use those lists to create word search puzzles.

Title: *Wordsearch Creator*
Publisher: Centron

Enter your own word list and generate a word search puzzle instantly.

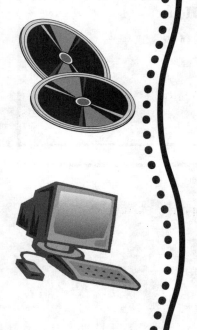

REFERENCE SOFTWARE

Title: *3-D Dinosaur Adventure*

Publisher: Knowledge Adventure

Equipped with three-dimensional glasses, you and your students explore a prehistoric world populated by the most chilling and thrilling creatures that ever lived.

Title: *A Field Trip to the Rainforest*

Publisher: Sunburst

The food chain activity challenges students to identify predator/prey relationships.

Title: *Animal Classifications*

Publisher: Centre Communications

This is an exciting, new way for students to learn about four classifications of animals.

Title: *Animal Kingdom*

Publisher: Unicorn

It introduces students to the wonders of the animal kingdom and various zoological species.

Title: *Animals*

Publisher: Clearvue

Carlos and Henry, under the guidance of Maggie, a wise old owl, lead students on an exploratory tour of animals, fish, amphibians, insects, birds, and mammals.

Title: *Basic Electricity*

Publisher: Quality Visions

It introduces the concept of current flow and open, closed, and short circuits. It also stimulates logical thinking.

Title: *Dinosaur Days*

Publisher: Toucan

It displays environments, including plants and animals that actually existed during that era. Using hundreds of prehistoric parts, create realistic or fanciful creatures that ride skateboards or munch on plants.

Title: *Dinosaur Kids*

Publisher: Nordic

Cage Velociraptors in a game of dots. Leave no Dino Rock unturned in a game of concentration or play SoliDinoTaire. There are seven games.

REFERENCE SOFTWARE (cont.)

Title: *Earth Science*
Publisher: Clearvue

Rocks and minerals, volcanic eruptions, the oceans, mountains, deserts, the air and atmosphere, rain and clouds, the water cycle, and more are illustrated and discussed.

Title: *Firefighter*
Publisher: Simon Schuster

Students will experience the thrill of being firefighters while learning valuable information about what to do in the case of a real fire.

Title: *Five Senses*
Publisher: SVE

This excellent introduction to our "windows on the world" celebrates the importance of our senses.

Title: *Jimmy Saves the Day!*
Publisher: Science for Kids

This is a 30-page original story about a lad named Jimmy whose unique ability to change into the shapes of simple machines is called upon frequently by a hapless llama named Dolly.

Title: *Junior Nature Guides: Insects*
Publisher: ICE

Learn about hundreds of insect species and their habitats.

Title: *Kids and the Environment*
Publisher: Tom Snyder

Role-play the captain of the school's soccer team. Time is short, the team needs to practice, and the field is covered with litter. Should they pick up the litter or practice anyway?

Title: *Learn About Dinosaurs*
Publisher: Sunburst

Students learn the habitats, sizes, food, and prey of nine dinosaurs. They dig up and put together a dinosaur fossil. They can recreate on-screen the environment in which the dinosaurs lived and write and print stories about it.

REFERENCE SOFTWARE (cont.)

Title: *Multimedia Bug Book*
Publisher: Expert Software

Dr. Anson Pantz's bug collection has escaped, and he needs your help to retrieve them. There are five different habitats to explore.

Title: *Plants*
Publisher: Clearvue

Let your students explore plants and their natural environments with the help of Mother Nature and a rather large and intelligent bee.

Title: *Safety for Children: Playground Safety*
Publisher: Churchill Media

A group of children demonstrate safe play on outdoor equipment at both school and public playgrounds.

Title: *Sammy's Science House*
Publisher: Edmark

In five fascinating activities, students learn about plants, animals, and seasons and weather and sort plants, animals, and minerals.

Title: *School House Science*
Publisher: Sierra

Students will learn about topics such as natural resources, our planet's weather, kinds of energy, life cycles of living beings, the elementary forces of electricity, and much more.

Title: *Science Blaster Jr.*
Publisher: Davidson

As they explore the Blaster Ship's high-tech, fun-filled science lab, students will actively participate in experiments as they discover the joys and the wonders of science.

Title: *Science II*
Publisher: Advanced Learning Systems

It builds science foundations, including topics such as rocks and minerals, the solar system, weather, water, matter, machines, electricity, computers, heat, light, cells, plants, insects, the food chain, and more.

Title: *Water Planet*
Publisher: Science for Kids

Explore the earth and its exciting water cycle with Winston the water molecule and his pals.

SLIDE SHOW STORYBOARD 1

Slide # _____

Words/Narration _____
_____
_____
_____
_____

Slide # _____

Words/Narration _____
_____
_____
_____
_____

Slide # _____

Words/Narration _____
_____
_____
_____
_____

Slide # _____

Words/Narration _____
_____
_____
_____
_____

SLIDE SHOW STORYBOARD 2

Slide # _____

Slide # _____

Slide # _____

Slide # _____

© Teacher Created Materials, Inc.   143   #2427 Integrating Technology into the Science Curriculum

BIBLIOGRAPHY

Barron, Ann E. and Gary W. Orwig. *New Technologies for Education—A Beginner's Guide.* Libraries Unlimited, 1995.

Bennett, Steve and Ruth. *The Official Kid Pix Activity Book.* Random House, 1993.

Chan, Barbara J. *Kid Pix Around the World—A Multicultural Activity Book.* Addison Wesley, 1993.

Cowan, Bill. *Computer Basics.* Teacher Created Materials, 1995.

Gardner, Paul. *Internet for Teachers and Parents.* Teacher Created Materials, 1996.

Garfield, Gary M. and Suzanne McDonough. *Creating a Technologically Literate Classroom.* Teacher Created Materials, 1995.

Haag, Tim. *Internet for Kids.* Teacher Created Materials, 1996.

Hayes, Deborah. *Managing Technology in the Classroom.* Teacher Created Materials, 1995.

Healey, Deborah. *Something to Do on Tuesday.* Athelstan, 1995.

Lifter, Marsha. *Integrating Technology into The Curriculum* (Primary). Teacher Created Materials, 1997.

Lifter, Marsha. *Writing and Desktop Publishing on the Computer* (Primary). Teacher Created Materials, 1996.

Pereira, Linda. *Computers Don't Byte.* Teacher Created Materials, 1996.

Pereira, Linda. *Computers Don't Byte* (Primary). Teacher Created Materials, 1996.

Reidel, Joan. *The Integrated Technology Classroom—Building Self-Reliant Learners.* Allyn & Bacon, 1995.

Willing, Kathleen R. and Suzanne Girard. *Learning Together—Computer Integrated Classrooms.* Pembroke Publishers Ltd., 1990.

Wilson, Jim and Dave Youngs. "Colorful Currents." AIMS, September, 1995: 7–9.

Wodaski, Ron. *Absolute Beginner's Guide to Multimedia.* Sams Publishing, 1994.

ONLINE SERVICES

America Online, (800) 827-6364

Classroom PRODIGY Service, (800) 776-3449, ext. 629

CompuServ, (800) 848-8990

Earthlink, (800) 395-8425

Netscape Navigator, (415) 254-190